Mysteries of Galaxy Formation

Françoise Combes

Mysteries of Galaxy Formation

 Springer

Published in association with
Praxis Publishing
Chichester, UK

Dr Françoise Combes
Observatoire de Paris
LERMA
Paris
France

Original French edition: *Mystères de la formation des galaxies*
Published by © Dunod, Paris 2008

Ouvrage publié avec le concours du Ministère français chargé de la culture – Centre national du livre
This work has been published with the help of the French Ministère de la Culture – Centre National du Livre

Translator: Bob Mizon, 38 The Vineries, Colehill, Wimbourne, Dorset, UK

SPRINGER–PRAXIS BOOKS IN POPULAR ASTRONOMY
SUBJECT *ADVISORY EDITOR*: John Mason, M.B.E., B.Sc., M.Sc., Ph.D.

ISBN 978-1-4419-0867-4 Springer Berlin Heidelberg New York

Springer is a part of Springer Science + Business Media (*springer.com*)

Library of Congress Control Number: 2009936157

Apart from any fair dealing for the purposes of research or private study, or criticism or review, as permitted under the Copyright, Designs and Patents Act 1988, this publication may only be reproduced, stored or transmitted, in any form or by any means, with the prior permission in writing of the publishers, or in the case of reprographic reproduction in accordance with the terms of licences issued by the Copyright Licensing Agency. Enquiries concerning reproduction outside those terms should be sent to the publishers.

© Copyright, 2010 Praxis Publishing Ltd., Chichester, UK

The use of general descriptive names, registered names, trademarks, etc. in this publication does not imply, even in the absence of a specific statement, that such names are exempt from the relevant protective laws and regulations and therefore free for general use.

Cover design: Jim Wilkie
Translation editor: Dr John Mason
Typesetting: BookEns Ltd, Royston, Herts., UK

Printed in Germany on acid-free paper

Contents

	Preface	ix
	List of illustrations	xi
1	**Going back in time to observe the young universe**	1
	The telescope: a time machine	1
	The horizon of our universe	3
	Horizon and expansion of the universe	6
	The far universe, at various distances	7
	Olbers' Paradox	9
	The development of structures	12
	The formation of galaxies requires the existence of exotic matter	13
	But how do structures of different sizes collapse?	15
	The evolution of the galaxies: a 'live report'	17
	Blue galaxies galore, at high redshift	21
	A surprising inversion of scale	23
	Astronomers, the archaeologists of galaxies	24
	Where do stellar halos come from?	25
	Galaxy archaeology: the Milky Way	*29*
2	**Infant galaxies in their cocoons**	37
	Searching for distant galaxies	37
	Large Lyman-α mapping projects	40
	Energy distribution within a galaxy	43
	The nature of dust	44
	Large molecules act as small dust grains	45
	More or less dusty galaxies	47
	A means of detecting remote galaxies: millimeter waves	47
	The results of millimeter-wave research	49
	The perturbed morphology of the first galaxies	51
	The beginning of the story...	56
	Questions still unanswered	60
3	**The origins of black holes**	67
	What is a black hole?	67
	Do black holes of the 'galactic' type exist?	69
	Black holes and galaxies	72
	How many black holes are there in the universe?	72

vi **Mysteries of Galaxy Formation**

	How does a black hole grow?	78
	The first black holes in the early universe, and intermediate-mass black holes	79
	Binary black holes and the possibility of observing them	82
	The observation of binary black holes: clues to the demography of black holes?	84
	Activity in black holes: 'downsizing'	85
	Self-regulation phenomena	87
	And what if the opposite were true?	89
	In conclusion…	91
4	**Scenarios of galaxy formation**	**93**
	The formation of structures: 'top-down' or 'bottom-up'?	93
	The formation of structures by mergers	96
	Several scenarios for galaxies	99
	The secular evolution of galaxies	106
	Environmental effects	108
	Bimodality between red and blue galaxies	112
	Feedback between supermassive black holes and star formation	116
	Dwarf elliptical or dwarf spheroidal galaxies	116
5	**The problem of dark matter**	**121**
	The large-scale structure of the universe: the success of the CDM (Cold Dark Matter) model	121
	Baryonic oscillations: another success for the CDM model	124
	Did visible matter follow dark matter? The bias	126
	Dark matter and the scaling relations of galaxies: the Tully-Fisher law for spirals	129
	Dark matter and the fundamental plane of elliptical galaxies	132
	Has the ratio of dark matter to visible matter evolved over time?	136
	The first major problem with the CDM model: cusps	137
	The second major problem with the CDM model: angular momentum	139
	The third major problem with the CDM model: satellite halos	139
	So what is dark matter?	143
6	**How can the problems be solved, and with what instruments?**	**149**
	Successes and problems: where are we now?	149
	Dark-matter particles: self-interacting or colliding?	150
	A first avenue of approach: a better understanding of complex baryonic processes	152
	Modified gravity	156
	The MOND problem in clusters of galaxies	158
	MOND and the formation of galaxies	161
	The modification of matter, not gravity	162

... and how does string theory come into this? 163
Instruments of the future: ALMA, JWST, ELT, SKA... 164

Glossary 175

Appendices 181

Index 185

Preface

The universe around us is composed of galaxies. Some occur in groups of about ten, some in clusters of hundreds, others in superclusters. How did all these structures come to be? Where did they come from?

Let us take our own Milky Way galaxy as an example. It appears to us as a pale 'milky' band arching across the sky, its glow emanating from a multitude of stars. Our Sun is one star among the hundreds of billions that populate the Milky Way.

A galaxy is a collection of stars (typically about one hundred billion), with gas and dust forming the interstellar medium from which new stars are born. The mystery of the formation and evolution of galaxies is complex, and involves knowledge of many ideas and phenomena concerning the birth of the universe. These will become apparent as we journey through this book.

In Chapter 1, we present the context within which all these events occur: the expansion of the universe from the Big Bang onwards, and the first 'inhomogeneities' from which arose the early structures of the universe. This is the framework enclosing all the 'characters' or celestial bodies as they evolve, and it is essential first of all to set the grand scene, even though it will be reworked, and details will be added, as we proceed.

When we talk of the universe, we are staggered by the distances and timescales involved. We shall encounter extraordinary orders of size: the 150-million-kilometer distance between the Earth and the Sun already seems large to us, and light takes a little over eight minutes to traverse it from our familiar local star. However, this 'astronomical unit' is too small to serve as a yardstick, and we use light-years; one light-year being the distance light travels in one year at 300,000 kilometers per second. Now, the region of the universe which we are going to describe is more than ten billion light-years across: so an even bigger unit, the *parsec*, comes into its own. A parsec is equivalent to 3.26 light-years (or about 3.10^{13} km), or 200,000 times the distance between the Earth and the Sun!

Then we shall see how, contrary to our everyday experience, space and time appear intimately connected. In astronomy, a telescope is a time machine, as we see in Chapter 1 and beyond. We can look today at the past of distant galaxies and, although we describe them in the present tense, we are really describing their youthful stages, forever lost in the past. This mixing of timescales can be somewhat unsettling at first, but we soon become used to it.

Each chapter is preceded by a brief summary. This gives us an idea of the general content of the chapter, and of the terms and notions defined within it. The chapters are not necessarily meant to be read in a linear, continuous manner, and the summaries will allow readers to move from one chapter to another and back again according to their own particular logic.

Although technical terms are defined when they first appear, there is a Glossary available, which also defines terms used in the book.

Françoise Combes
August 2009

Illustrations

1.1*	The Andromeda Galaxy (M31).	2
1.2	A wide-angle view of the Virgo Cluster of galaxies.	4
1.3	Schematic representation of the horizon as a sphere around a given point in the universe.	5
1.4	The difference between luminosity distance and angular diameter distance as a function of redshift.	8
1.5	Artist's impression showing the Wilkinson Microwave Anisotropy Probe.	10
1.6	Anisotropies in the cosmic microwave background.	12
1.7	When a structure forms, it must first of all decouple from the expansion.	14
1.8	The history of the formation of structures in the universe.	16
1.9	Edwin Hubble's name was given to the space telescope launched in April 1990.	18
1.10*	Several thousand distant galaxies in a long-exposure image (Hubble Deep Field North).	19
1.11	Small segment from the Hubble Ultra Deep Field (HUDF).	20
1.12	Energy distribution for a range of galaxies.	22
1.13	Evolution in the rate of star formation in the disk of our Galaxy.	23
1.14	History of stellar formation in some Local Group galaxies.	26
1.15	Our neighbor, the Andromeda Galaxy, in a somewhat unfamiliar guise.	28
1.16	M32, the small elliptical companion of the Andromeda Galaxy.	29
1.17*	An artist's impression of the very early universe.	31
1.18	Relation between metallicity and age of stars in our Galaxy.	32
1.19	Abundance ratio between alpha-elements and iron, for stars in the solar neighborhood.	33
1.20*	A near-infrared view of our Galaxy.	35
2.1	The history of star formation in the universe.	38
2.2	Model of the shell ejected by starburst activity at the center of a galaxy.	42
2.3	The distribution of energy in the spectra of typical spiral galaxies.	46
2.4	Redshift in the spectra of starburst galaxies.	48
2.5	Deep-sky survey in the millimeter domain of an area behind a nearby cluster of galaxies.	50
2.6	The 'cloverleaf' quasar.	52
2.7*	The Hubble Ultra Deep Field (HUDF).	53

xii **Mysteries of Galaxy Formation**

2.8	Hubble Tuning Fork diagram.	54
2.9	Clumpy galaxies.	56
2.10*	Simulations of clumpy galaxies.	57
2.11	Candidates for the title of the most remote galaxies currently known.	58
2.12	Distant galaxies on a small region of the Hubble Ultra Deep Field (HUDF).	61
2.13	Evolution of galaxy merger rates.	62
2.14	Emission in the Lyman-á line from a large area of neutral hydrogen gas.	63
2.15	Searching for high-redshift candidates (at $z > 7$), with the 'gravitational telescope.'	64
3.1*	Accretion disk around a black hole in a binary star system.	70
3.2.	The 'ballet' of stars at the center of our Galaxy.	71
3.3*	The quasar 3C273 in the constellation of Virgo, and its host galaxy.	73
3.4*	Hubble Space Telescope face-on view of the small spiral galaxy NGC 7742.	74
3.5	Relationship between the mass of the black hole and the mass of the bulge of a galaxy.	75
3.6*	Black-hole-powered jet of sub-atomic particles streaming out from the center of the galaxy M87.	76
3.7	Evolution in the number of quasars throughout the history of the universe.	77
3.8	Schematic representation of the re-ionization era of the universe.	81
3.9*	Double radio source, two pairs of jets and a binary black hole in the process of formation.	83
3.10	Light curve of the quasar OJ 287, in visible light.	84
3.11	Distribution in redshift of the 23,000 quasars in the 2dF catalogue.	86
3.12	Luminosity function of quasars.	88
3.13*	Self-regulation phenomena in NGC 1275 (Perseus A), the central galaxy of the Perseus cluster.	90
4.1*	Three-dimensional survey of the large-scale structures of the universe.	95
4.2	Great walls of galaxies in cross-section.	96
4.3*	A slice through a small part of the simulated universe, showing the large-scale structure in the dark matter.	97
4.4	Schematic representation of a merger tree.	98
4.5*	The Antennae galaxies.	100
4.6*	A pair of interacting galaxies nicknamed 'The Mice.'	101
4.7	Representation of the Milky Way, seen edge-on.	102
4.8*	M80, one of the densest of the known globular star clusters in the Milky Way.	103
4.9*	Artist's impression of our Galaxy, surrounded by a stream of stars.	104
4.10	The two main scenarios of galaxy formation: monolithic collapse and hierarchical formation through mergers.	105
4.11	The secular evolution scenario.	106

4.12	The possible scenarios of galactic evolution, combining the hierarchical and the secular evolution theories.	108
4.13*	The Coma Cluster of galaxies.	109
4.14*	Abell 1689, a massive cluster of galaxies that shows signs of merging activity.	110
4.15	Bimodality between 'blue' and 'red' galaxy sequences.	113
4.16	The most massive galaxies have high surface density, while dwarf galaxies have low surface density.	114
4.17*	One of the newly discovered low-luminosity dwarf spheroidal galaxies, accompanying the Milky Way.	118
4.18	The distribution of the dwarf satellite galaxies of the Milky Way and the Andromeda Galaxy.	119
5.1	Amplitude of fluctuations as a function of size.	123
5.2	Sound waves in the cosmic microwave background and in the distribution of galaxies.	125
5.3	Gravitational shear.	127
5.4	Illustration of the Integrated Sachs-Wolfe (ISW) effect.	129
5.5	Tully-Fisher relation for spiral galaxies.	131
5.6	Scaling relations for elliptical galaxies.	133
5.7	Hubble Space Telescope image of the Helix Nebula.	135
5.8	The cusp problem, predicted in simulations, when observing galactic cores.	138
5.9	Tully-Fisher relation obtained in numerical simulations of dark matter (CDM).	140
5.10	Simulations of structures in the CDM model.	141
5.11	The problem of the missing satellite galaxies.	142
5.12	The Large Hadron Collider (LHC).	146
6.1	Numerical simulations of dark-matter halos.	151
6.2*	Starburst activity in the 'Cigar galaxy,' M82.	153
6.3	The influence of the active nucleus on the dynamics of the host galaxy.	155
6.4	Rotational velocity of our Milky Way galaxy as a function of distance from the center.	157
6.5	Rotation curves of a selection of galaxies of different types.	159
6.6*	Composite image of a collision between two clusters of galaxies.	160
6.7	Schematic representation of a 'brane' model of the universe.	164
6.8	Artist's impression of the SuperNova Acceleration Probe (SNAP).	166
6.9	Artist's impressions of the Atacama Large Millimeter/submillimeter Array (ALMA).	168
6.10	Artist's impression of the E-ELT (European Extremely Large Telescope).	169
6.11	Computer-generated illustration of the Thirty Meter Telescope (TMT).	170
6.12	Artist's impression of the James Webb Space Telescope (JWST).	171
6.13	Artist's impression of the central area of the Square Kilometer Array (SKA).	172

xiv **Mysteries of Galaxy Formation**

A.1 Components of the universe. 183
A.2* The principal stages in cosmic history. 184

(* Figures marked with an asterisk also appear in the color section positioned between pages 118 and 119.)

1 Going back in time to observe the young Universe

The huge advances made during the last few decades in our understanding of the formation of galaxies is due to the ever increasing quality of telescopes. Modern instruments are capable of detecting very distant galaxies, thereby reaching through time, across about 95 percent of the age of the universe. What then is the volume of the universe now accessible to us? There is a natural limit: that of the observable horizon, towards which we are looking back today.

In order to chart this accessible volume, we must define distances: not an easy thing to do in an expanding universe. There are several different scales of distance, a phenomenon to which we are unaccustomed in our own locality of the universe.

How were the first structures formed? In the initial 'soup' of ionized particles, dark matter and photons, primordial density fluctuations (the seeds of present-day structures) are visible to us today in the form of (extremely weak) anisotropies in the cosmic microwave background. Will there be time for them to collapse under the effect of gravity, in spite of the expansion of the universe?

High-definition images from the Hubble Space Telescope allow us not only to follow this evolution directly, by observing distant galaxies, but also to observe individual stars in nearby galaxies, with a view to discovering their respective ages, and reconstructing the evolution of the galaxies as we look back though time. This 'archeological study' is done preferentially in our own Galaxy, the Milky Way, where the chemical tagging of the stellar populations can also constrain their speed of formation.

The telescope: a time machine

Who has not one day dreamed of travelling back through the ages, of living life as our great-grandparents knew it, of being at the court of King Henry VIII, or seeing the Enlightenment as it happened?

In a way, the telescope provides a means of going back in time, but with one proviso: we also have to travel through space. The further into space we look, the

2 **Mysteries of Galaxy Formation**

Figure 1.1 The Andromeda Galaxy (M31) is the closest large spiral galaxy to our own Milky Way. Located 2.5 million light-years away one can easily locate it with the naked eye in the constellation of Andromeda on clear, moonless nights. This overlapping three frame mosaic shows the galaxy and its two small satellites M32 (above center) and NGC 205 (below left of center). The dynamic range of the combined image has been compressed significantly to show the inner and outer regions of the galaxy (Adam Block, NOAO, AURA, NSF). See also PLATE 1 in the color section.

Going back in time to observe the young Universe

further back into the past we see, and the younger the galaxies appear. So we cannot see how our own Galaxy, the Milky Way, formed and lived its early life, but we can see the formation of very distant galaxies. This magical time machine works because of the finite speed of light (approximately 300,000 kilometers per second), which no form of radiation can exceed, whatever the velocity of its source. No signal emitted by nearby galaxies travels faster than this.

On a local scale, the view we have of our own Solar System is not instantaneous, since light takes several hours to reach us from its most distant objects. Indeed, it was this phenomenon which led to a first estimate of the speed of light by the Danish astronomer, Ole Rømer, working at the Paris Observatory in 1676. The Andromeda Galaxy (Figure 1.1) is a near neighbor, at a distance of about two-and-a-half million light-years[1]. We do not therefore have an 'up-to-date' view of its spiral arms, but see their configuration as it was some two-and-a-half million years ago. We contemplate the Virgo Cluster (Figure 1.2), the nearest large cluster of galaxies to our own, and see it as it was 65 million years in the past: when dinosaurs still lived on Earth.

The further we look out into space, the further back in time we go. The images we now obtain of the most distant galaxies are formed by light which they emitted about 13 billion years ago, when the universe was only 5 percent of its present age! It seems likely that, today, these galaxies at the edge of the observable universe will have evolved considerably, or even merged with neighboring galaxies. If we could see them as they are today, we would not recognize them!

The horizon of our universe

Each point in the universe (which is perhaps infinite, a question to which we shall return) is at the center of a sphere representing the universe observable from that point. Around our Galaxy, the Milky Way, there is just such a limiting horizon. The radius of this sphere is the distance light has travelled since the beginning of the universe, the Big Bang. Since we know accurately the age of the universe, 13.7 billion years, the distance to this horizon is 13.7 billion light-years.

Even if many galaxies exist beyond that horizon, we cannot see them: their signals would take longer than the age of the universe to reach us. Such considerations give us insight into the realities of space-time. We could of course imagine galaxies similar to our own, at the same stage of their evolution, stretching out to infinity: but that is not what we actually see, because as we approach our limiting horizon, we approach the epoch of the Big Bang.

---

[1] A light-year is a unit of distance. It is the distance that light can travel in one year. Light moves at a velocity of about 300,000 kilometers per second. So in one year, it can travel almost 10 million million km. More precisely, one light-year is equal to 9,460,528,400,000 or 9.4605284×10^{12} km.

4 Mysteries of Galaxy Formation

Figure 1.2 A wide-angle view of the Virgo Cluster of galaxies. It contains more than a hundred galaxies of many types, several of which are among the brightest galaxies visible from Earth. The large mass of the cluster, which includes not only the visible galaxies, but also large amounts of tenuous, hot gas visible only at X-ray wavelengths, and even larger amounts of 'dark' matter, is pulling our Galaxy and the other galaxies in our Local Group of galaxies toward the Virgo Cluster. Thus, in a sense, we may be thought of as an outlying part of the Virgo 'supercluster' (Digitized Sky Survey, Palomar Observatory, STScI).

Going back in time to observe the young Universe 5

However, some of the galaxies which we see as young systems will certainly seem more evolved as seen from the center of the observable universe of astronomers located in other galaxies, far away across the universe – and they will see galaxies invisible to us (as we see some that are invisible to them)! As Figure 1.3 shows, galaxies at various stages of evolution are scattered before our

Figure 1.3 Schematic representation of the horizon as a sphere around a given point in the universe. The observer is at the center of the sphere. Its radius is the distance travelled by light in 13.7 billion years, the time that has elapsed since the Big Bang. Observing objects at great distances amounts to travelling back in time: the observer sees galaxies as they were when they emitted the light which has just arrived. Currently, we have been able to look back across 95 percent of the age of the universe. The edge of the sphere corresponds to the Big Bang. Soon after the Big Bang, the universe consisted of charged particles: a plasma opaque to light rays diffused by ions and electrons. Here, this phase is represented by the opaque dappled ring. About 380,000 years after the Big Bang, ions recombined to form hydrogen atoms, and the 'dark age' of the universe (the dark part of the sphere) began, lasting until the first galaxies appeared. When today's observers detect the photons of the cosmic microwave background, the vestiges of the Big Bang, they are looking back through time to the 'surface of last scattering' of the photons, i.e. the inner (opaque) surface of the dappled band on the diagram.

eyes: near the horizon, they appear in their formative stage at the frontier of the 'dark age' of the universe, which we shall now describe. To read the open book of evolution, we need only observe in great depth: to great distances.

Although any observer sees only a part of the universe, a second observer in another galaxy will see other objects invisible to the first. Each observer is surrounded by a personal horizon of visibility.

Horizon and expansion of the universe

The horizon of the universe also evolves over time. Firstly, it increases as the universe ages, since its radius is the distance travelled by light since the Big Bang. In order to know if there are more galaxies, we must take the expansion of the universe into account.

The expansion of the universe was discovered in the late 1920s by Edwin Hubble, in company with Vesto Slipher and Milton Humason. They noticed that, with the exception of the nearest systems, the redshifts of galaxies are proportional to their distances. This led to the famous 'Hubble's Law,' relating recession velocity v and distance d:

$$v = H_o d$$

where H_o is a constant of proportionality known as the 'Hubble constant.' Redshift is often interpreted as a Doppler effect, according to which the frequency of the radiation emitted by a receding (or approaching) object is lower (or higher) than its rest frequency. In the case of sound waves, the Doppler effect is familiar to us: for example, the pitch of the siren of an approaching fire engine sounds higher than when it is moving away from us. Since the light (or more specifically the spectral lines) of distant galaxies is shifted towards the red, the intuitive interpretation is that those galaxies are receding from us, and the further away they are, the faster they recede.

But this recession of the galaxies is only apparent. In reality, the expansion of the universe corresponds to an increase in all distances. We can take the example of a balloon being inflated: imagine that the universe has two dimensions and is on the surface of the balloon. The galaxies are dots marked on that surface. As the balloon expands, all the galaxies will move apart. None is in a privileged location: none is at the center of the universe. Each one 'sees' every other galaxy receding at a speed proportional to its original distance. These are not true movements, and the analogy with the real Doppler effect breaks down when the redshift is well above 1, and galaxies are apparently receding at more than the speed of light.

To what is the redshift due, according to this interpretation? Simply put, the wavelength of the light emitted is also elongated due to the expansion, as is every other distance. The further the light travels, the 'redder' it becomes, i.e. its wavelength is stretched by the expansion of the universe. So the redshift of distant galaxies is greater. Let us take the example of a spectral line, emitted at a reference wavelength of λ_0. On its journey, the wavelength of the photon will have lengthened to wavelength λ as received by the observer. From this we define the redshift z as $(\lambda - \lambda_0)/\lambda_0$. This redshift will increase as the source of radiation recedes, and the expansion has more time to do its work. The relationship between wavelengths is also equivalent to that of the characteristic scales of the universe between the times of emission (t) and reception (t_0 = today), i.e. λ/λ_0 = R (t_0)/R (t) = 1 + z. We see how redshift can be used to measure distances and times in the universe. This redshift is intimately linked to the law of expansion, represented by the dimensionless scale factor R (t), which by convention assumes a value of zero at the time of the Big Bang, increasing to a value of R (t) = 1 today.

Let us now return to the definition of our observable horizon. How can the number of galaxies within that horizon increase, if the galaxies are receding as the distance to our horizon grows, simply because of the amount of time that has passed since the Big Bang? Which motion will prevail: on the one hand, our horizon moving further away as time passes, or on the other, the galaxies receding more and more and crossing beyond that horizon? It is not possible at present to answer this question, since the law governing the expansion of the universe is not monotonic.

To know how the horizon varies and how many galaxies cross it in either direction, we would have to consider in detail the whole history of the expansion of the universe and the result will depend on the law of the variation in time of the scale factor R (t). In practically all cosmological models, the expansion is initially very rapid, and then slows. But what happens subsequently is largely dependent upon the content of the universe.

In certain models of the finite universe, where the expansion slows and is even reversed, it is possible that all the galaxies are visible, and that there is no horizon. On the other hand, it seems from current observations that we are in a universe whose expansion is accelerating, but the speed of the horizon is nevertheless greater than the expansion, and more and more galaxies will become visible in the near future. This may, however, not be the case in a more distant future.

The far universe, at various distances

There are several ways to determine the distances of objects in nearby space, either by measuring their apparent size and comparing it with their actual size, or by measuring their apparent brightness and relating it to a standard intrinsic brightness. All these methods converge, and give similar results, in the case of the nearby universe. However, as soon as we move further out, typically for galaxies with redshifts greater than 1, all these definitions cease to be equivalent, and several distance scales can be defined for the same object. Two of these distance scales are of essential importance for observations of the oldest known galaxies, dating from the birth of the universe:

8 Mysteries of Galaxy Formation

- angular diameter distance, which is based on the fact that the angular size of an object is inversely proportional to its distance;
- luminosity distance, which is based on the fact that the apparent brightness of an object is inversely proportional to the square of its distance.

Einstein's Theory of General Relativity predicts a luminosity distance which is far greater than angular diameter distance (Figure 1.4). So galaxies further and

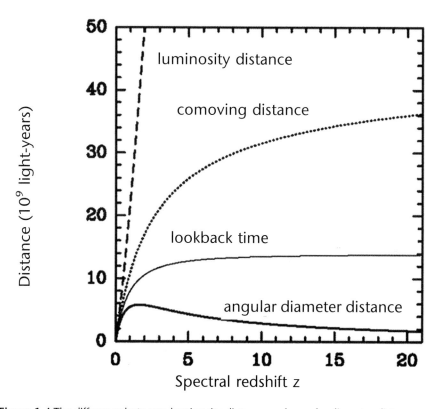

Figure 1.4 The difference between luminosity distance and angular diameter distance as a function of redshift. Outside the context of the nearby universe, there are several definitions of distance which are not coincident. Luminosity distance is defined in terms of the relationship between the absolute (intrinsic) magnitude of an astronomical object and its apparent magnitude (the latter decreases as the square of its luminosity distance). The angular diameter distance is defined in terms of the intrinsic size of an object and the apparent size (which decreases with the luminosity distance). While a body decreases in brightness as its redshift increases, its apparent size decreases much less. In a way, the universe acts as a gravitational lens, making the most distant objects appear larger. On the graph above are shown: the comoving distance, involving a correction for the expansion of the universe; distance expressed as time (13.7 billion years since the Big Bang, with the Hubble constant $H_0 = 70$ km/s/Mpc, the quantity of dark energy $\Lambda = 0.73$ and the total quantity of matter $\Omega_m = 0.27$ (see text).

further away according to their redshifts maintain a reasonable angular size accessible by telescopes (of the order of a second of arc), though they appear less and less bright and difficult to detect. The ratio of these two distances is $(1 + z)^2$, and can reach more than 100 for z = 10.

To try to subtract the mechanical effect due to expansion in the variation of the distance between two objects, the idea of their 'comoving' distance is introduced, telling us where the galaxies are now. Thus, if the two objects are not going to fall into each other, but are merely receding with the expansion, their comoving distance is constant. The comoving distance, corrected for expansion, is the angular distance multiplied by $(1 + z)$.

It is also possible to define distance by time elapsed based upon the age of the universe. Of course, in the nearby universe ($z \ll 1$) all distance scales are equivalent, which is why the phenomenon of multiple distances is not intuitive.

We therefore perceive that the detection of high-redshift galaxies will be difficult, since their brightness decreases as the square of the luminosity distance, or as $(1 + z)^4$ times the square of the angular distance, which is itself almost constant.

Olbers' Paradox

It will be of interest to dwell for a moment on the paradox which is usually attributed to the German astronomer Heinrich Wilhelm Olbers, who drew attention to it in the 1820s although he was not the first to pose the problem: why is the sky dark at night? If the universe is infinite, then the light of the galaxies should make the sky shine equally brightly in all directions.

Nowadays, it is easy to see how the paradox can be resolved. The combination of the finite speed of light and the finite character of the universe in time, starting with the Big Bang 13.7 billion years ago, implies that we see only those galaxies situated within the horizon. Also, as the universe expands, the light from the most distant galaxies is shifted towards the red, at frequencies different from those of nearby galaxies. So, to each domain of wavelengths (or to each color) there corresponds a finite tranche of the universe: the night sky is not bright because the light that we see is never the sum of the light from an infinite number of galaxies.

Although the sky is not bright at any wavelength, it is brighter in certain colors, and study of the cosmic microwave background is instructive at all wavelength domains. The 'brightest' domain is of course that of millimeter waves, where vestigial photons from the Big Bang are observed. These correspond to black-body radiation at a temperature of 2.725 Kelvin (or about –270 degrees Celsius).

Initial fluctuations
The first structures (and by 'structures' we mean collections of matter in the process of formation, such as galaxies, clusters of galaxies, and clusters of

10 Mysteries of Galaxy Formation

Figure 1.5 Artist's impression showing the Wilkinson Microwave Anisotropy Probe WMAP in position at the L2 Lagrange point, 1.5 million km beyond the Earth (NASA, WMAP Science Team).

clusters) condensed as a result of density fluctuations in matter and radiation. These are now better understood thanks to observations of the cosmic microwave background, the fossil black-body radiation which is the echo of the Big Bang.

In 2002, the American WMAP (Wilkinson Microwave Anisotropy Probe) satellite (Figure 1.5) followed up the work of NASA's COBE, and other ground-based and balloon-borne instruments, with large-scale studies of the anisotropies in the microwave background. WMAP's spatial resolution was much higher than that of its predecessors. This cosmic microwave background radiation (CMBR) has a thermal black body spectrum which peaks in the microwave range at a frequency of 160.2 GHz, corresponding to a wavelength of 1.9 mm, similar to that inside the microwave ovens in our kitchens. The CMBR dominates the sky in its particular frequency domain, and it is remarkably homogeneous and isotropic: this tells us that the universe must have been extremely homogeneous just after the Big Bang.

In order to be able to study these faint primordial fluctuations of the universe, like ripples on the surface of the otherwise homogeneous background, astronomers must first subtract several components of greater amplitude. The first subtraction is that of a constant corresponding to the value of the mean background. Then, a dipolar component: one half of the sky appears blue (cooler) and the other red (warmer) – merely a manifestation of our own motion compared with the background radiation, the radiation that is, in a way, the absolute frame of reference for the universe. Indeed, our Galaxy moves relative to the ensemble of the large-scale structures, at a velocity of the order of 600 km/s in the direction of the so called 'Great Attractor' (a very massive cluster of galaxies), and we detect the associated Doppler effect in the observed frequency of photons. Once the dipolar component is subtracted, we must also take into account the emission of our own Galaxy at these wavelengths, which now becomes visible. Subtracting it is made easier by the fact that its spectral signature is not that of a black body, and its distribution in space is not homogeneous. After all these steps, it becomes possible to reveal the small temperature fluctuations of the cosmic background, of the order of 1 in 100,000. It is these anisotropies which give us information about the formation of structures.

Figure 1.6 charts these fluctuations in the universe. What we see there corresponds to the surface of last scattering, dating back to 380,000 years after the Big Bang. In the beginning, the universe was very hot and dominated by radiation; matter was ionized, in a plasma of protons and electrons interacting closely with photons and scattering them. The universe was opaque. When the expansion of the universe cooled it to a temperature of about 3,000 Kelvin, the protons and electrons recombined to form hydrogen atoms, and the universe became neutral. Photons were no longer scattered by charged particles, and now moved in straight lines. The universe became transparent.

Observing this radiation today, cooled to a temperature of 2.725 K, we look back upon that opaque surface. At this time, fluctuations in density in which matter and photons participated were stable. They did not collapse under the effect of their own gravity, and therefore correspond to waves moving through the medium. The term 'sound waves' here refers to the fact that photons and matter participate in these vibrations like a gas traversed by sound.

The characteristic scales of the maxima and minima of these oscillations tell us about the nature of the matter and their observed angular size today within the geometry of the universe. The fundamental mode of the waves corresponds to the size of the sound horizon at that epoch, which is a known dimension. The comparison with the apparent dimension today, which is of one degree of arc in the sky, shows that the photons have moved in a straight line and that the universe has no curvature – it is flat. The other modes of oscillation (harmonics), their positions and their amplitudes depend on the quantities of ordinary (baryonic) matter and exotic matter, and their degree of dissipation (damping).

12 Mysteries of Galaxy Formation

Figure 1.6 Anisotropies in the cosmic microwave background. This chart represents the whole sky as seen by the WMAP satellite, designed to observe the sky background at millimetric wavelengths. The anisotropies are observed on the surface of last scattering, 380,000 years after the Big Bang. They represent very small fluctuations in that background, of the order of one part in 100,000. In order to detect them, it was necessary first to subtract the continuous and first-order homogeneous and isotropic background emission, and then the dipole, corresponding to our own motion relative to the cosmic microwave background (or the absolute frame of reference of the universe). Finally, account had to be taken of the foreground radiation of our own Galaxy at these wavelengths. The fluctuations are the vestiges of the primordial perturbations which gave rise to the great structures of the universe and to galaxies. They manifest themselves as temperature variations within a range of 200 microKelvin (NASA, WMAP Science Team).

The study of these peaks in the spatial distribution of the fluctuations in radiation therefore gives much information about our universe.

The development of structures

The principal 'motor' at the origin of the formation of structures is gravity relayed by instabilities which will cause structures to collapse under the effect of their own gravity. We are accustomed to thinking that the formation of stars has its roots in gravitational instability; this is certainly very efficient, for as soon as a volume of gas has reached a critical mass, density increases exponentially, the cloud collapses and in free-fall time, or nearly so, a star is born. In the case of structures in the universe, this is not so easily accomplished: expansion works against self-gravity, all the while the structure is not gravitationally bound.

It is normal to define the comoving frame of reference, which dispenses with the question of expansion: in this system, all lengths and distances are measured with reference to a 'meter' which stretches with the expansion. The length of this 'meter' is 1 today and was 0 at the time of the Big Bang, the characteristic (dimensionless) size R/t defined above. It is possible to show that, in this comoving frame of reference, density fluctuations grow linearly (and not exponentially), and their rate of growth is proportional to the rate of expansion. This phase of slow growth occurs while the contrast in density within the fluctuation is low, relative to the mean density. As soon as the density within the fluctuation becomes twice as great as the mean density, evolution becomes non-linear and collapse can proceed.

We can then consider this small region of the universe as decoupled from expansion. The size of this evolving 'structure' is at one moment in expansion, more slowly than the universe, then reaches a maximum radius before contracting and rebounding, with much dampened oscillations (Figure 1.7). The maximum radius of the structure is twice its radius of equilibrium. Indeed, the evolution of this structure is similar to the condensation of the universe having the critical density to collapse inwards upon itself.

The formation of galaxies requires the existence of exotic matter

The growth of structures is so slow that we might ask ourselves how the universe can be old enough for the galaxies we see today to have formed. Let us imagine that the universe consisted of only baryonic matter, i.e. protons, neutrons and all the atoms made from them. At the beginning of the universe, this matter was ionized, and charged particles were very strongly coupled with photons through electromagnetic interaction. As long as the photons could not collapse into structures, matter would not collapse and would remain with the photons. Density fluctuations could not therefore develop before the recombination of the initial plasma into hydrogen atoms, which occurred 380,000 years after the Big Bang.

In other words, the ions were closely coupled with photons, and these, with their strong pressure, hindered gravitational collapse. Only when baryonic matter decoupled from photons, at the epoch of the recombination, was it able to collapse. When did this recombination occur? Hydrogen atoms recombine at a temperature of about 3,000 K. So, the question arises: when did the temperature of the universe reach 3,000 K? Shortly after the Big Bang, the temperature of radiation was many millions of degrees, then the universe progressively cooled as it expanded, until the temperature of the cosmological black body reached its current value of almost 3 K. Just as white-hot iron emits light at a shorter wavelength than that emitted by red-hot iron, which is at a lower temperature,

14 Mysteries of Galaxy Formation

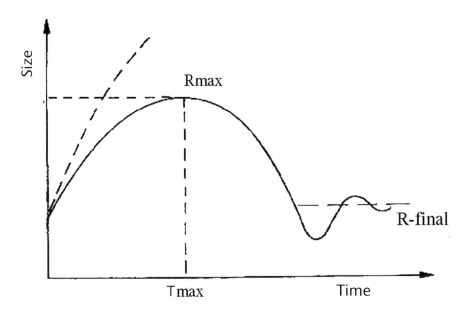

Figure 1.7 When a structure forms, it must first of all decouple from the expansion. Here we see the radius around a fixed mass M, which will become a distinct structure. Time elapses towards the right. At first, the mass continues to expand (solid line), at a rate which slows in relation to the mean rate of the expansion of the universe (dashed line). The relative density becomes greater and greater relative to the rest of the universe. The time will come (T_{max}) when the density reaches the critical density that allows the structure to collapse upon itself. Then the expansion will be reversed until the structure reaches virial equilibrium, and kinetic energy is balanced by potential energy. During the collapse, the velocities of agitation are higher, until the equivalent 'pressure' balances out the forces of gravity. We then obtain a secular stable structure, of radius R_{final}.

the temperature of a black body is inversely proportional to the characteristic wavelength it emits. In redshift terms, wavelength increases as the inverse of $(1 + z)$. Similarly, the temperature of the universe decreases as $T_0(1 + z)$, the temperature of the cosmological black body today if $z = 0$.

Since the current temperature is of the order of 3 K, the recombination corresponds to a redshift (z) of about 1,000. This factor of 1,000 is also the factor of the expansion of the universe since the recombination, since it is strictly equal to the redshift factor. We now come up against a problem as far as the growth of structures is concerned: their amplitude can increase only by a factor of 1,000 since the recombination. Now, measurements from WMAP and other cosmic-background experiments have revealed amplitudes 100 times smaller, of the order of 1/100,000. In order for the amplitude of density contrasts to become of the order of unity, and for structures to be able to decouple from expansion and form galaxies, a further growth factor of 100 would be required! We must therefore conclude that ordinary, baryonic matter is not enough!

Another kind of matter, more exotic and non-baryonic, is definitely needed, with particles that do not interact with photons. Fluctuations in this matter can therefore begin their growth well before the recombination of ordinary matter, and attain today the necessary amplitude, of the order of unity. The fluctuations of this hypothetical matter, which emits no radiation and is therefore called dark matter, will begin to grow as soon as the gravitational influence of the photons upon this matter becomes negligible; this happens when the densities of the photons and the matter reach equivalence. In fact, the density of matter in the expansion varies as $1/R^3$, where R is the characteristic size of the universe. The energy density of the photons decreases faster, as $1/R^4$ because the energy of each photon (proportional to its frequency ν or inversely proportional to its wavelength λ) decreases with the expansion: the wavelength is redshifted as $1/R$, and the number of photons decreases as $1/R^3$.

Even if photons dominate initially, matter will therefore eventually prevail as soon as the redshift becomes 10 to 100 times greater than that of the recombination. From this epoch of the equivalence of matter and radiation onwards, fluctuations in the dark matter will increase, and will already have formed gravitational wells into which baryons will collapse at the time of the recombination, and will make up the lost ground. So we can say that galaxies, or at least embryonic galaxies, indeed formed before the stars which would go on to shine in these gravitational wells after the recombination.

But how do structures of different sizes collapse?

Certainly, there exist all sorts of structures of different sizes, often involved with each other, such that future galaxies will be included in future clusters of galaxies, etc. The distribution in amplitude of fluctuations as a function of their size, known as the spectrum of fluctuations, depends on the supposed theory at the origin of primordial fluctuations. However, this theory is still little understood. Currently, it seems that an inflationary phase of the universe is required, but there exist various theories of inflation. All these theories predict a spectrum of fluctuation sizes which has no characteristic scale, given that the law of gravity has no preferential scale. The favored distribution of amplitudes today is that in which fluctuations all enter the horizon with amplitudes independent of size (equal to 3/100,000 in relative intensity). Before the epoch of the equivalence of matter and radiation, fluctuations could grow only if they were larger than the horizon. When the horizon caught them up, the influence of the photons and their pressure prevented all growth. This phenomenon breaks the similarity of the various scales, favoring the large ones.

At the time of the recombination, the first structures to become gravitationally unstable were those that were just above critical mass, i.e. one million solar masses. This would be the equivalent of today's globular star clusters or very

16 Mysteries of Galaxy Formation

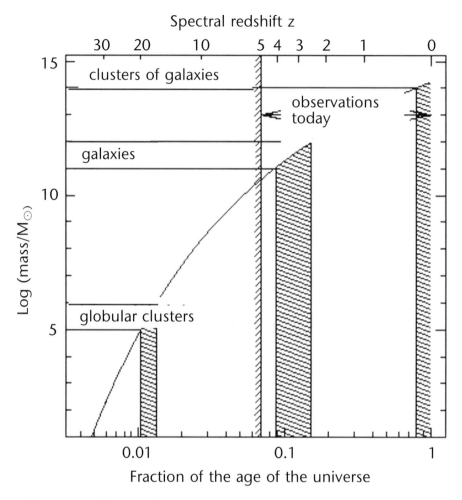

Figure 1.8 The history of the formation of structures in the universe. Small structures form first, and the largest structures last. Just after the recombination era of the universe, the first structures to collapse under their own gravity are of the size of globular clusters (about one million solar masses). Then small structures merge to become larger ones, according to the scenario of hierarchical formation. Increasingly massive galaxies form, then groups of galaxies, which coalesce to become clusters of galaxies. The largest structures, clusters of galaxies essentially at $z = 1$, form later, or even today in the case of the most massive. Finally, superclusters are forming today, and their formation will continue into the near future.

small dwarf galaxies. Then, gradually, larger and larger structures became unstable and decoupled from the expansion, as shown in Figure 1.8. More massive structures can be considered to have formed by mergers of smaller structures included in their volume. This is the hierarchical theory of the formation of structures. But the existence of clusters of galaxies is certainly in some way already defined before the collapse of smaller structures, since the fluctuations are all initially in place.

Hierarchical formation is an essential mechanism in the growth and formation of galaxies. Numerical simulations of a set of universes, beginning with initial cosmological conditions, and reflecting the fluctuations expected after recombination, show how small structures form in cosmic filaments, and merge. These successive mergers can be quite easily reproduced by analytical calculation, if only dark non-baryonic matter, subject only to the forces of gravity, is considered. In effect, we can avail ourselves of the fact that the force of gravity exhibits the same behavior on all scales, and that there is no favored scale. This means that we have a kind of symmetry of scale, and can expect that the mass distribution of galaxies will follow a power law independent of scale, and self-similar. This is indeed what is observed in the spectrum of galaxies: this spectrum varies as a power law, until a maximum is reached, beyond which the number of galaxies falls off exponentially in a kind of sudden cut-off.

About 35 years ago, two American astronomers, William H. Press and Paul L. Schechter, from the California Institute of Technology (CalTech), studied the self-similar development of structures under the influence of successive mergers. They showed that the final result has little to do with the initial conditions: after several successive stages of mergers, a self-similar equilibrium was established and the number of galaxy mergers bringing the mass to a category M equaled the number of mergers which took the galaxies M into a higher category M+dm. So this mass spectrum is defined in a universal way. This was a considerably successful approach.

The evolution of the galaxies: a 'live report'

The Hubble Space Telescope (Figure 1.9) has led to enormous advances in the detection and identification of very distant objects, thanks to its resolution, about 0.1 second of arc, in the visible domain. This means that we can now see images of distant galaxies. The major problem for ground-based telescopes is the presence of the Earth's atmosphere, whose turbulence blurs the images as air masses move ceaselessly between the telescope and the celestial bodies. This diffraction and refraction of light rays by air masses of different temperatures and densities, with different refractive indices, limits telescopes to a resolution of typically one second of arc or worse. So not only is it impossible to see any detail in smaller-scale images, but also the detection of distant objects becomes much more difficult since their light is spread, instead of being concentrated at one point in the image.

Although the Hubble Space Telescope has only a 2.4-m mirror, from its vantage point 610 km above the Earth it has been able to detect far more galaxies than an equivalent telescope could have detected from the ground. We need only to look at the Hubble Deep Field North images of galaxies in Figure 1.10:

18 Mysteries of Galaxy Formation

Figure 1.9 Edwin Hubble's name was given to the space telescope, eventually launched in April 1990, and seen here in orbit about 600 km above the Earth, whose rich harvest would revolutionize astrophysics in all its aspects (NASA).

although the field is not a large one, just a few square arcminutes, the number of galaxies detected broke all records: 3,000 galaxies, and most of them very distant!

As we can see in Figure 1.11, which is a detailed view taken from the Hubble Ultra Deep Field, the very distant galaxies appear to be very irregular in shape, and some objects look like collections of 'grains,' which may represent the 'building blocks' from which galaxies formed. In Figure 1.11, for example, there are eighteen large collections of stars, so close to each other that they will doubtless merge in the near future, in about 100 million years. They perhaps show us the way in which galaxies form. It is very difficult to 'see' a galaxy in formation, because, unlike the formation of a star, the formation of a galaxy is not a rapid affair, but rather a succession of events stretching over billions of years.

The 'building blocks' observed by Hubble are seen at such a huge redshift that they correspond to the epoch when the universe was only a billion years old:

Going back in time to observe the young Universe   19

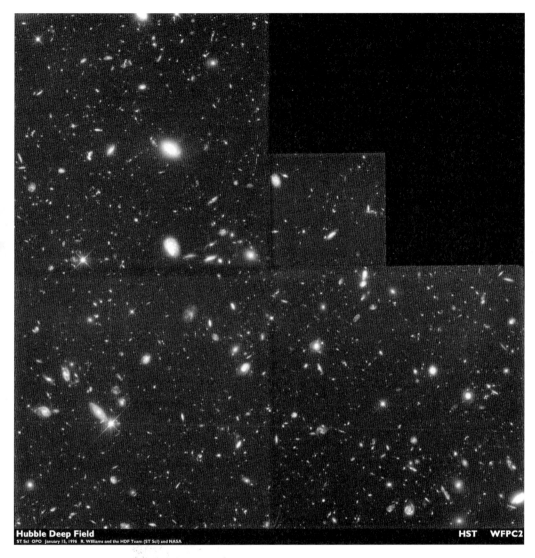

Figure 1.10 Several thousand distant galaxies in a long-exposure image (Hubble Deep Field North). An image from the Hubble Space Telescope of a small area in the northern hemisphere of the sky. This region was observed in 1996, and this 10-day exposure was built up from 342 separate images. Although the region of sky shown is only 2.5 minutes of arc across, more than 3,000 galaxies can be identified, thanks to the high sensitivity and good quality of the image (the resolution being 0.1 second of arc). This image was taken during a 'Director's discretionary time' slot (Robert Williams was Director of the Space Telescope Science Institute at this time). It was published as soon as it was obtained, so that ground-based spectroscopic research could be concentrated on this region, chosen because there are so few foreground objects in our own Galaxy (R. Williams (STScI), the Hubble Deep Field Team and NASA). See also PLATE 2 in the color section.

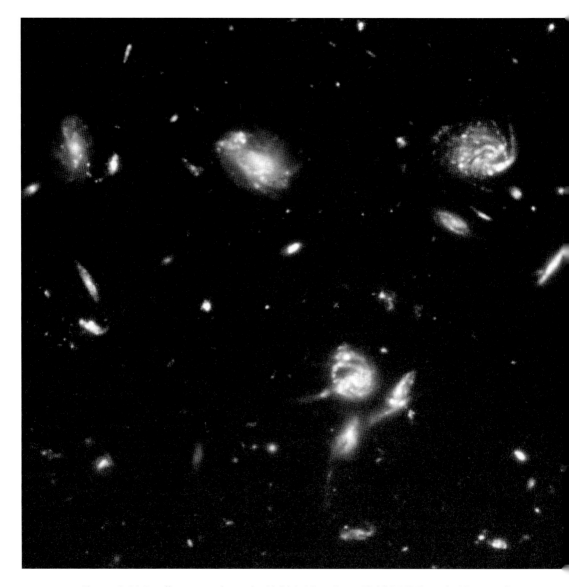

Figure 1.11 Small segment from the Hubble Ultra Deep Field (HUDF), only 30 seconds of arc wide, of an area free of foreground objects in the southern hemisphere of the sky. The original HUDF image, acquired by instruments on board the Hubble Space Telescope in 2004, is 3.3 × 3.3 minutes of arc in size, and the million-second (or 11-day-long) exposure of the region reveals about 10,000 galaxies. Several filters were used in combination to construct the original true-color image (NASA, ESA, S. Beckwith (STScI) and the HUDF Team).

7 percent of its current age. These objects are far smaller than galaxies, but rather similar in size to a small galactic bulge, about 10 percent of the diameter of the disk of the Milky Way. The dominant color of the galaxies is also different as a function of redshift, which shows a considerable temporal evolution in the rate of star formation. When galaxies form a large number of young stars, their color becomes very blue, unless those stars are obscured and reddened by dust.

How can we determine the redshift of these remote galaxies? The quantity of light emitted is too faint for the Hubble Space Telescope to create a spectrum of these objects. We therefore have recourse to larger ground-based telescopes, 10 m in diameter, but the objects are mostly too small and too faint. Therefore, the photometric method is used. It is only an approximate method, and consists in using the characteristic shape of energy distribution as a function of wavelength for a given galaxy. This distribution is not flat, but exhibits many peculiarities, humps, troughs and peaks, which allow us to identify the wavelength of the emission. By comparison with the wavelength received, the redshift and therefore the distance of the galaxy may be determined.

One of the most characteristic humps in the spectrum of a galaxy is the Lyman break, which corresponds to an ionization limit in a hydrogen atom (912 ångström, or 912 Å[2]). The most energetic (bluer) photons will be able to ionize the H atom, but below this threshold the atom will remain neutral. Hydrogen atoms are the most abundant in the universe, and the further away a galaxy is, the more hydrogen atoms there are along the line of sight between us and the galaxy. Now all these hydrogen atoms along this line of sight, whether they belong to a galaxy or just to a cosmic filament, will absorb photons of greater (but not lower) energy. The Lyman break serves as a criterion to identify certain distant galaxies, known as Lyman break galaxies. More generally the breaks occur closer and closer to the infrared or visible domains as galaxies are more and more remote. There is also a characteristic break at 4,000 Å due to several lines in the atmospheres of old stars, which allow us to achieve certain identifications. If the photometry is carried out using large numbers of measurements and color filters, identification is easier and more certain (Figure 1.12).

Blue galaxies galore, at high redshift

The first astronomers to study the numbers of galaxies in Hubble Deep Field images were surprised: blue galaxies are relatively ten times more numerous, which suggests a considerable temporal evolution. By contrast, galaxies considered today as 'evolved' (high-mass ellipticals and lenticulars, lacking gas and with very little star formation) do not seem to be evolving, as if they had formed very early on in the universe, while large spirals like our own

---

[2] The ångström (symbol Å) is an internationally recognized non-SI unit of length equal to 0.1 nanometer or 1×10^{-10} m.

22 Mysteries of Galaxy Formation

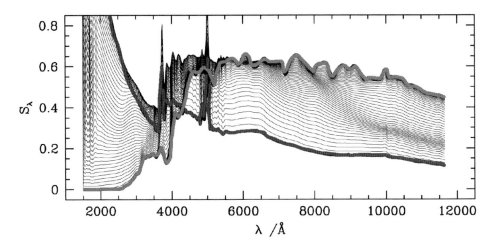

Figure 1.12 Energy distribution for a range of galaxies. Each curve represents the total flux of a galaxy as a function of the wavelength in ångströms (from ultraviolet on the left to infrared on the right). The red and blue curves show the extreme energy distribution of a red or a blue galaxy respectively (after Csabai *et al.*, 2003).

appear to have formed stars at an almost constant rate throughout their lives (Figure 1.13).

These galaxy counts were soon accomplished, since they took no account of distances (redshifts), but rather concentrated on the number of galaxies as a function of their luminosity and morphological type (especially spiral and elliptical). A number of interpretations arose. This excess of faint blue galaxies could be due to a great number of dwarf galaxies in the near past. However, once redshifts were obtained, this interpretation no longer seemed viable. Alternatively, it is possible that there were many early mergers of galaxies, given that such interaction favors star formation and would therefore tend to make the galaxies look blue? However, the products of these mergers, corresponding to elliptical or lenticular galaxies, should also be evolving as a consequence, and this is not observed. There were indeed a greater number of interactions between galaxies in the past, but this seems to have evolved more slowly than the number of blue galaxies, as shown by the number of galaxy pairs as a function of redshift, z.

So we must be cautious in our interpretations, because there are a number of other factors that can cloud our view:

- an evolution in numbers can be confused with an evolution in mere luminosity;
- the morphological types of galaxies can appear abnormally irregular since, as we go back in time, we do not see remote galaxies in the same colors as modern galaxies. This effect is caused by the expansion of the universe, and the stretching of the wavelengths along the light path. In particular, an

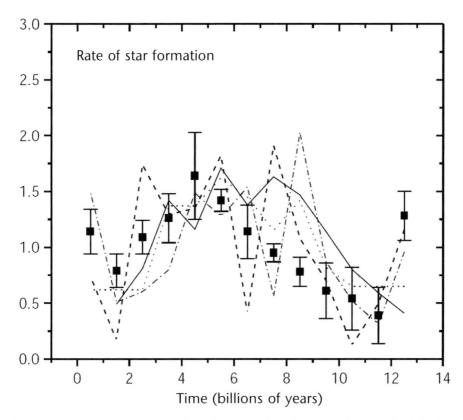

Figure 1.13 Evolution in the rate of star formation (in solar masses/year) in the disk of our Galaxy, the Milky Way, normalized to mean rate, as determined by various methods. Contrary to the exponentially decreasing law that we might expect if the Galaxy were progressively using up the gas present since its inception, star formation continues at an almost constant rate, having slowed slightly during the last few billion years and then recently increased. This constantly renewed rate of star formation implies an almost continuous accretion of gas (after Rocha-Pinto and Maciel, 1997).

image observed in visible light in fact reveals the morphology of the galaxies in the ultraviolet, the wavelength of emission. This wavelength tends to favor irregular sites of the formation of young rather than old stars, though the latter in fact constitute most of the mass.

A surprising inversion of scale

One of the important characteristics of the evolution of galaxies is the apparent temporal inversion of scales; many observations have shown that giant galaxies completed their evolution early on in the universe, and certainly during the first half of its lifetime, while the evolution of smaller galaxies, and dwarf galaxies, was proceeding later. Could this observation run counter to the predictions of the 'hierarchical formation' theory? *A priori*, we might expect the reverse to be true, since large galaxies form through mergers of smaller ones.

On closer inspection, we see that things are not quite so simple. The hierarchical formation theory applies first and foremost to the majority of matter, i.e. non-baryonic dark matter in galaxies and large structures. As far as this matter is concerned, nothing proves that the largest structures were not formed later, by mergers of smaller structures. On the contrary, we know that clusters of galaxies are 'younger' than groups and galaxies, and that superclusters of galaxies form only today. These large structures can merge without their component galaxies merging. Indirectly, observations of clusters of galaxies show that the dark-matter halos of galaxies have merged into one common halo, but the baryonic structures of the galaxies themselves have remained distinct. The tracers of dark matter in clusters are, on the one hand, X-rays which reveal the hot gas in hydrodynamic equilibrium in the gravity well of the cluster, and gravitational lenses, enabling us to trace out the lines of the total gravitational field.

So it seems that hierarchical formation is indeed a characteristic of dark-matter halos. But what of ordinary, baryonic matter? Here, the phenomena are much more complex, and for the most part poorly understood. We can imagine that the gas which will form stars can pass through various phases, very hot or very cold, and that its loss of energy and angular momentum, necessary for the formation and evolution of a galaxy, depend on a number of interrelated factors. It is possible that the earliest star formation heated the gas to such a degree that further condensation and star formation were prevented. Gas may even have been completely ejected from the galaxy if its gravitational well was not deep enough, as is the case with dwarf galaxies.

Moreover, the environment of a galaxy has a considerable effect upon its evolution: interactions with nearby galaxies can favor star formation, or inhibit it if they are too rapid. In groups and clusters of galaxies, gas can be heated by interactions, and become stabilized as hot gas, emitting X-rays, in the intracluster medium between the galaxies. The motions of the galaxies through this medium give rise to an extragalactic wind which sweeps away their interstellar gas, and star formation ceases. Galaxies are no longer supplied with cool gas and their evolution is halted, which could explain the lack of evolution observed in massive elliptical and lenticular galaxies, while star formation may proceed today in dwarfs. This does not mean that large galaxies are not formed by mergers of smaller galaxies, earlier in the universe.

Astronomers, the 'archaeologists' of galaxies

Astronomers have for a long time tried to reconstruct the history of the formation of our Galaxy from the study of its various stellar populations. They have distinguished, in general, two categories: the young population of stars in the disk, and the older population on the halo. The different populations are distinguished not only by age, but also by their chemical composition, spatial distribution and kinematic properties. This will be explored further later in this chapter.

In our Galaxy, and, thanks nowadays to the Hubble Space Telescope, in the nearby galaxies of the Local Group, it is possible to study stars one by one. Their spectra give information about their ages and chemical abundances. The latter is a tracer of the metallicity of the gas that formed the star, the gas being a product of nucleosynthesis in earlier stars that ejected it into the interstellar medium at the end of their lives. The abundance of gas can also be diluted by events involving the accretion of gas, lacking in heavy elements, from outside the galaxy. In the most distant galaxies, stars cannot be observed individually, and this kind of study has to be carried out through averaged quantities along the line of sight, which may involve stars of different populations. So it is necessary to carry out a synthesis of stellar populations and compare the result to the observations. This is often a tricky problem, since there is no unique solution, and it is difficult to work out which is the right one from a number of possibilities.

By these 'archaeological' means, it has been possible to retrace the history of star formation in our Galaxy. One important thing we have realized is that the rate of star formation has remained remarkable constant through time, across billions of years, with the exception of a few local fluctuations. This does not fit at all well with the predictions that might be made for an isolated galaxy, in which the quantity of gas, and by extension the rate of star formation, must decrease exponentially as time goes on, over a typical period of three billion years. So the galaxy must have received gas from outside, at a well sustained rate, throughout its evolution. This fact is also corroborated by observations of the abundances of elements. In particular, the relationship between age and metallicity, and the relative abundances of iron and oxygen, show us that the importation into the galaxy of gas that is not much enriched is required.

The observation by the Hubble Space Telescope of a large number of stars, and their color-magnitude diagram, has enabled us to trace the history of star formation in nearby galaxies of the Local Group. Figure 1.14 shows some of these 'histories.' For these galaxies, individual stars can be identified and placed onto a diagram with a view to separating out the different populations, according to their color and apparent luminosity. For our neighbor the Andromeda Galaxy, the stellar populations vary enormously as a function of the region observed, evidence of the recent very violent and disturbed past of this giant galaxy.

Where do stellar halos come from?

In the 1960s, the theory put forward to explain the formation of our Galaxy and its halo of old stars involved the monolithic (i.e. occurring as a single event)

26 Mysteries of Galaxy Formation

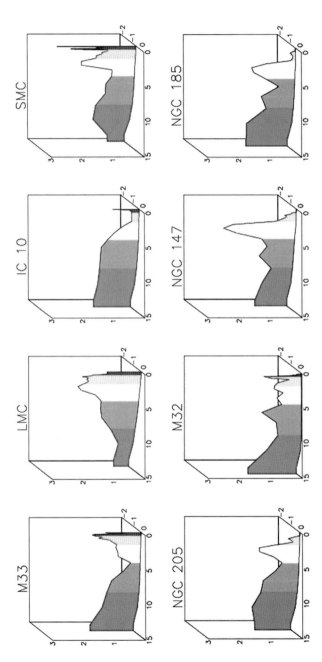

Figure 1.14 History of stellar formation in some Local Group galaxies. Each diagram corresponds to a galaxy, and their designations are shown above the diagrams. The vertical axis represents the rate of star formation, normalized to the mean rate (as in the previous figure), and the horizontal axis represents the regression into the past relative to the present time ($t = 0$, on the right). The third axis (depth) represents the metallicity of corresponding stars (relative to the Sun). The colors represent types of stars (old stars in red on the left, young stars in blue on the right) and their metallicity increases with time. In effect, the heavy elements are formed by nucleosynthesis within stars, which throughout their lives have ejected enriched gas into the interstellar medium. New stars forming within this medium will have greater metallicity (only accretion of the insubstantial gas from outside can dilute the abundance of heavy elements). The first four galaxies (M33, LMC, IC10 and SMC) represent spiral or dwarfs still possessing gas today; the other four objects represent elliptical galaxies which are poorer in gas, and in which formation has ceased. Note that in all cases, star formation is not a function of exponentially decreasing time (after Dolphin et al., 2005).

collapse of a more or less spherical volume of gas. Since stars formed progressively during this collapse, the earliest stars still retained a spherical structure which went on to become the halo of old stars. The gas, flattening out more and more as its rotational velocity increased, conserving angular momentum, became the disk within which young stars were formed.

This scenario is now discredited, and for several reasons. On the one hand, the problem of the constant rate of formation of stars in our Galaxy shows that it is not a closed system, but that gas continues to fall, and to form the disk even today. On the other hand, the halo of old stars consists of various different groupings according to their kinematic properties and spatial distribution. These groupings resemble tidal streams, and they are left over from small companion galaxies that have been destroyed.

It may therefore be that the stellar halo was entirely formed after the rest of the Galaxy, from debris falling in from companion galaxies orbiting the Milky Way as satellites. The satellite galaxies that have interacted most recently with the Milky Way are the Sagittarius and Canis Major dwarf galaxies, both inferred within the last decade from the coherent streams of stars that they have left within the halo. These star streams are present at all galactic longitudes, and can be reproduced in numerical simulations: the conclusion is that the tidal forces of our Galaxy tear satellite galaxies to pieces when they come too close.

Our Galaxy has had no violent interactions with other giant galaxies in the last few billion years; if they had happened, it would be reflected in the rate of star formation, age distribution, and spatial disturbances due to tidal forces. The Andromeda Galaxy, however, has (recently) had a more violent past. Deep-field and wide-angle images of this galaxy show tidal streams, loops and disturbances characteristic of a recent merger with a relatively massive galaxy. Additional matter has increased the apparent size of the disk in the sky, as shown in Figure 1.15.

The question has even been asked as to whether the very particular structure of the Andromeda disk, where the spiral structure appears enveloped and dominated by a large ring of gas, dust and young stars, might not be caused by the passage through the disk of the companion galaxy Messier 32 (M32), which now is seen as an unusually shaped elliptical galaxy (Figure 1.16). The high-speed collision with M32 would have produced density waves within the Andromeda Galaxy, propagating from the center outwards, rather similar to the circular ripples produced in a pond by a stone thrown into it. In this collision, the smaller galaxy M32, perhaps originally a spiral, would have had its disk stripped away and would have retained only its bulge, thereby becoming a compact elliptical galaxy.

As we see, galaxies have been forming and evolving throughout the history of the universe; some are still forming today, or continue to evolve. Their destinies differ widely:

- some form very early on and very rapidly, then evolve passively as their stars gradually grow old, and no new stars are formed;
- other galaxies still exhibit bursts of star formation, and these galaxies are generally less luminous and less massive.

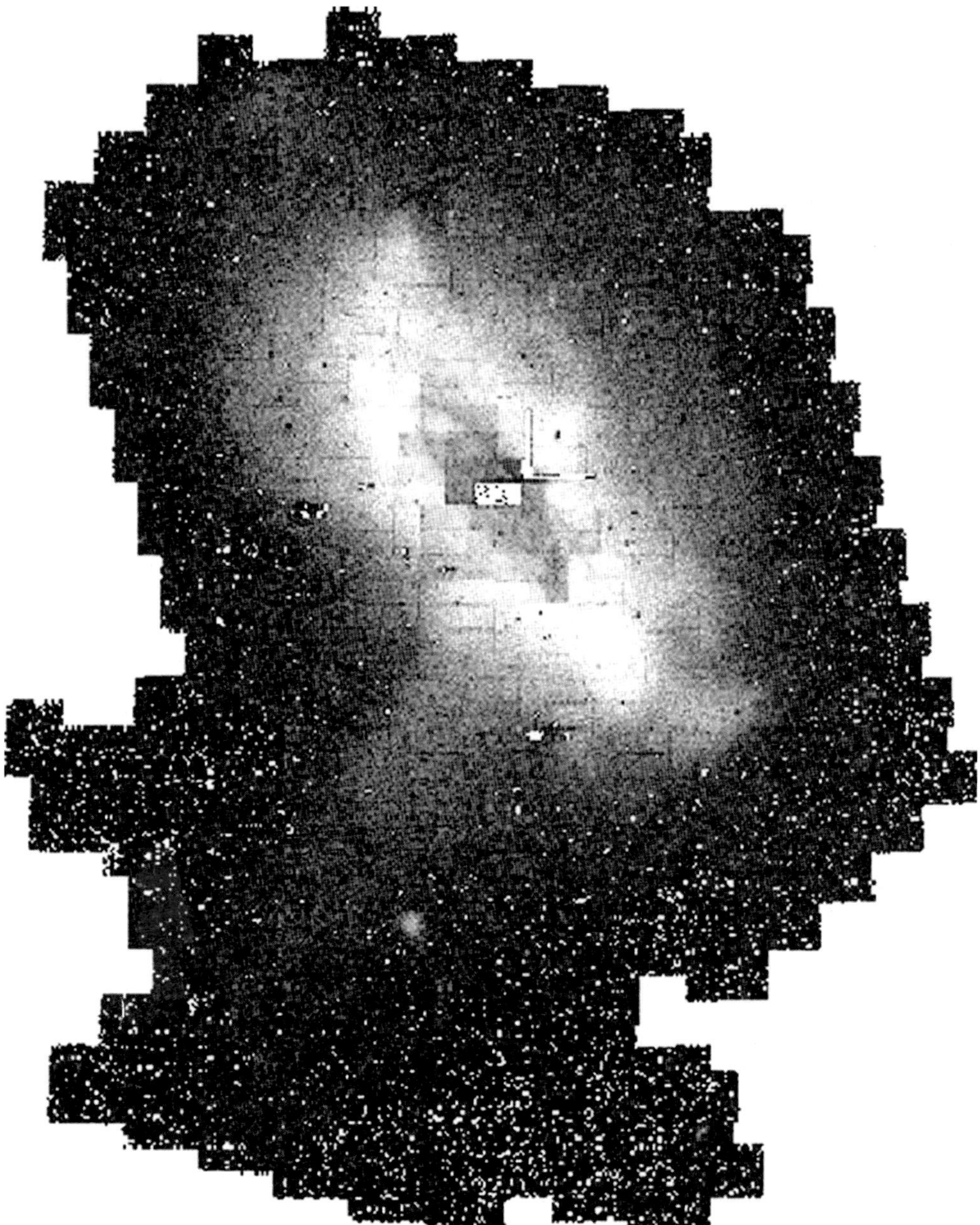

28 **Mysteries of Galaxy Formation**

Figure 1.15 Our neighbor, the Andromeda Galaxy, in a somewhat unfamiliar guise. The optical disk normally seen in photographs corresponds to the internal part. An integrated series of deep-field exposures shows that the disk is four times larger, and reveals tidal disturbances, loops and extrusions, suggesting that one or more companions have recently merged with this galaxy (after Ibata *et al.*, 2001).

Figure 1.16 M32, the small elliptical companion of the Andromeda Galaxy, is around 8,000 light-years across (longest dimension) and contains a few billion stars. Note that the background looks mottled in this image. This is due to the millions of unresolved stars in the disk of M31 which lies in the background (Adam Block, NOAO, AURA, NSF).

Can we understand the origin of these processes? Is it that we are seeing environmental effects, or are the fates of galaxies already sealed within the initial conditions of the universe?

Galaxy archaeology: the Milky Way

Stellar populations – the first stars
Our own Galaxy, the Milky Way, is a privileged place to practise galaxy archaeology, trying to reconstruct the history of the Galaxy from the age, metallicity and distribution of its stars. In the 1940s, different stellar populations were identified, in particular by German astronomer Walter Baade: the disk of our Galaxy was found to contain a population of young hot blue stars (or Population I), and the stellar halo to have an older and redder population of stars (Population II). Later it was discovered that the chemical composition varied

according to the populations, and that in general the youngest populations are more abundant in metals (in astronomy, all heavy elements – not just iron and the like, but carbon, nitrogen, oxygen, etc. – are called 'metals'). Their existence is due to the recycling of interstellar gas through several generations of stars, yielding metals through their internal nucleosynthesis.

To search for the oldest stars, and look back towards the formation of the Galaxy, it is then appropriate to search for the most metal-poor stars. These are however quite rare, certainly because most of the first stars have died. Indeed, it is widely believed that the first generation of stars (called Population III) formed just after the Big Bang from primordial gas (essentially hydrogen and helium) without the efficient cooling of metals. In these conditions, the fragmentation of the gas is less efficient, and only very massive stars can form – more massive than 100 solar masses. The lifetimes of massive stars are very short, of the order of a million years, and they explode as supernovae, enriching, very early on, the interstellar medium in metals (Figure 1.17). In the Galaxy, only two stars have been discovered with an abundance of iron [Fe/H][3] equal to 10^{-5} times the solar value, and about 200 with 10^{-4} the solar value. These first stars are quite precious in our quest to better understand the process of star formation in the early days: no Population III star has ever been observed, and we know very little of the gas fragmentation processes at these epochs.

There is also a good correlation between age, metallicity and the morphology and kinematics of the stellar components in the Galaxy. This is why there is hope that we can partly disentangle the formation of these components: the youngest component, the thin disk, is the most metallic, and also has the largest degree of rotation, of ordered motion. There is also a thick disk component, with more velocity dispersion in the stars, and made of older population stars, with lower metallicity. The stellar halo has almost no rotation, only velocity dispersion. The

---

[3] The metallicity of the Sun is about 1.6 percent by mass. For other stars, the metallicity is often expressed as '[Fe/H],' which represents the logarithm of the ratio of a star's iron abundance compared to that of the Sun. (Iron is not the most abundant heavy element, but it is among the easiest to measure with spectral data in the visible spectrum.) The formula for the logarithm is expressed thus:

$$[Fe/H] = \log_{10}\left(\frac{N_{Fe}}{N_H}\right)_{star} - \log_{10}\left(\frac{N_{Fe}}{N_H}\right)_{sun}$$

where N_{Fe} and N_H are the numbers of iron and hydrogen atoms per unit of volume, respectively. As will be seen from the formula, stars with a higher metallicity than the Sun have a positive logarithmic value, while those with a lower metallicity than the Sun have a negative value. The logarithm is based on powers of ten, so stars with a value of +1 have ten times the metallicity of the Sun (10^1). Conversely, those with a value of -1 have one tenth (10^{-1}), while those with -2 have a hundredth (10^{-2}), and so on. Population I stars have significantly higher iron-to-hydrogen ratios than the older Population II stars. Primordial Population III stars are estimated to have a metallicity of less than –5, that is less than one hundred thousandth of the abundance of iron which is found in the Sun.

Figure 1.17 An artist's impression of how the very early universe (less than one billion years old) might have looked when it went through the voracious onset of star formation. The most massive of these Population III stars self-detonated as supernovae, which exploded across the sky like a string of firecrackers (Adolf Schaller for STScI). See also PLATE 3 in the color section.

bulge is more mixed, containing both low- and high-metal stars, of various ages, and we will later discuss its possible origin. Globally, there is, as expected, a general trend of progressive enrichment of the Galaxy, as its stellar populations evolve and age, as displayed in Figure 1.18. The halo, bulge and disk occupy different regions in this diagram. Globally, the metallicity grew very quickly at the beginning of the universe, and continues to increase slowly with time.

There is also a striking point displayed in this figure, in that there are many more stars in the last five billion years than before. Does it translate to the history of star formation in the Galaxy? It is not certain, since the dating of stars can be done accurately only in a small volume in the solar neighborhood, and also, old

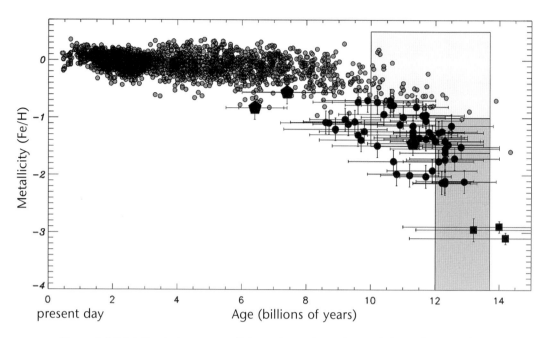

Figure 1.18 Relation between metallicity and age of stars in our Galaxy. There is a clear correlation, with older stars having lower metallicity, confirming an old rapid enrichment, followed by a slow progressive growth of metallicity. Note that the vertical scale is logarithmic and spans almost six orders of magnitude. The stars of the disk correspond to grey dots, while the black dots are the globular clusters. The three pentagons are for comparison globular clusters from the Sagittarius dwarf galaxy. The square symbols in the bottom right are halo stars, which are dated by radioactive elements with a long lifetime (such as uranium), detected in absorption in the stellar atmospheres. The rectangles represent the location of the bulge (light grey at top), and halo (dark grey at bottom) (from Haywood, 2009).

massive stars have died. Only stars of low mass can survive as long as the Hubble time. Also stars can migrate in the disk through the ages.

There are also precious diagnostics of galaxy evolution in the abundance ratios observed in each star. These abundance ratios are representative of the interstellar medium in which the star was formed, which itself corresponds to the enrichment history. The abundance ratios are a signature of the past speed of evolution and star formation. Stellar nucleosynthesis occurs in several steps. The first fusion reactions form helium (the alpha particle) from protons. Then, helium burning, through the 'alpha process,' or 'triple-alpha process,' forms the following more complex elements (called alpha elements) which are, among others: C, O, Ne, Mg, Si, etc… They are mainly synthesized by alpha-capture in the violent supernova explosions of massive stars (called Type II supernovae). This enrichment therefore happens very quickly. The other group of elements around iron, called the Fe elements, are formed in a much longer process. They

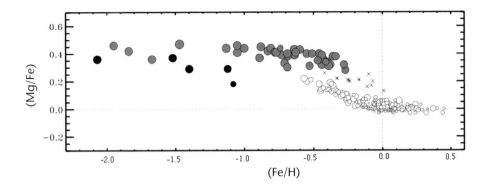

Figure 1.19 Abundance ratio between alpha-elements and iron, for stars in the solar neighborhood. The halo stars are represented by black dots, the thick disk stars by dark grey dots, and the thin disk stars by light grey smaller dots. Crosses are stars with properties intermediate between the two disks. All components have a different place in the diagram, showing that their stars have different formation scenarios (from Fuhrmann, 2003).

are released essentially in Type Ia supernovae, which are binary white dwarf explosions. These represent the end of the evolution of low-mass stars, and therefore take billions of years to form and enrich the medium. Iron is the element of lowest binding energy, and only neutron capture (either slow or rapid, s-process or r-process) can form heavier elements.

The most useful abundance ratio is then the alpha/iron element ratio, expressed as [α/Fe]. This ratio is enhanced when the star formation has proceeded in bursts, at a very rapid pace. This must have been the case in the early universe, where stars were more massive, forming more alpha elements relative to Fe. In general, high [α/Fe] ratio is found in massive elliptical galaxies, where stars appear to be formed faster than in disks of spiral galaxies. Figure 1.19 displays the abundance ratio between alpha-elements and iron, for stars in the solar neighborhood, according to which of the various stellar components of the Milky Way are involved. The abundance ratio plays a significant role in the chemical tagging of the various stars: the halo must have been formed in a rapid event at the beginning of the universe, while the stars of the disk have been formed in a quiescent manner, throughout the age of the universe.

Thin and thick disk
The principal component of the Milky Way is the thin disk. As for all spiral galaxies, its radial distribution is an exponential, i.e. the surface density decreases exponentially from the center to the border of the disk. It is the only component which possesses gas and still forms stars now. Apparently its star formation rate has remained quite constant during the last few billion years, and therefore there must exist radial flows of gas to replenish the disk. It is however difficult to separate different stellar populations in the disk, since there is continuously

radial mixing occurring in the disk, due to the bars and spiral arms propagating through it. The latter imply large-scale oscillations of the stellar orbits, and the mixing time is estimated at 1–2 billion years: and more rapid in the center, since the rotation period is shorter there. The spiral waves progressively heat the disk, by increasing the velocity dispersion at the expense of the rotation, and the younger stars can be traced through their dynamics.

The thin disk has a vertical scale of only one-tenth of its radius. Within this disk, older stars have a greater vertical dispersion, and greater depth, due to their more numerous scattering by spiral arms, or giant molecular clouds. However, there is a second component, clearly separated from this thin disk, with a height 3 or 4 times greater; this is called the thick disk. Its stellar population is older (similar to Population II), and constitutes a few percent of all the stars of the Galaxy. The thick disk is in rotation, but less than the thin disk, having in compensation more velocity dispersion. Its metallicity is intermediate between the metal-rich thin disk and the metal-poor halo.

There are at least two possible ways to form the thick disk, all related to galaxy interactions. The first and more likely one is that the Milky Way has in the past experienced a collision with a major perturber, which has heated and thickened the primordial disk. Then, the Galaxy would have continuously accreted gas, re-forming a thin disk. This scenario has the advantage of explaining the bimodality of the properties of the two disks, the stars of the thick disk all coming from the previous thin disk of the Galaxy. The second hypothesis is that interacting satellites would lose their stars during interactions, which will settle in a thick disk. Of course, the reality could be a mixture of these two possibilities. However, the abundance ratios observed for the thick disk (Figure 1.19) do not favor a multitude of small satellites, since the alpha/iron ratio is small in these dwarf galaxies. On the contrary, the first disk is likely to have formed stars more quickly, and given the radial mixture of the stars in the thin disk today, the observed abundances are compatible with the first hypothesis: a primordial thin disk thickened by a major perturber.

The peanut bulge

The near-infrared view of the Galaxy (Figure 1.20) reveals that our Galaxy has a very small bulge (normal for a late-type Sbc), which has the shape of a box or peanut. It is also triaxial, and elongated towards the observer, being part of the bar identified in the center. Chemically, the bulge is quite varied, containing the more metal-rich stars of the Galaxy, but also some of the most metal-poor. It contains in general old stars, but some are young. Its kinematics are intermediate between the disk and the halo, and it possesses some rotation. It is typically what is called a 'pseudo-bulge,' to distinguish it from the classical bulges that are similar to elliptical galaxies.

The most likely scenario for the formation of the peanut bulge is secular evolution due to the bar. Numerical simulations reproduce perfectly the elevation above the plane of the stars from the thin disk, when they resonate with the bar, to form a 'peanut' component. The abundance ratios show that the

Going back in time to observe the young Universe 35

Figure 1.20 A near-infrared view of our Galaxy. This panorama encompasses the entire sky as seen by the Two Micron All-Sky Survey (2MASS). The image is centered on the core of the Milky Way, toward the constellation of Sagittarius, and reveals that our Galaxy has a very small bulge (normal for a late-type Sbc galaxy), which has the shape of a box or peanut (University of Massachusetts, IR Processing & Analysis Center, CalTech, NASA). See also PLATE 4 in the color section.

bulge is relatively rich in alpha elements with respect to iron, compared to the disk population near the Sun. This means that the previous thin disk, from which stars have been derived to form the bar, was also richer in alpha elements than the present disk. This is compatible with what we have also seen for the thick disk.

The stellar halo: tidal streams

As was previously described, there is now a good deal of evidence that a lot of stars in the halo come from the infall of satellites. In particular, many tidal streams have been observed in the Milky Way halo: the most important one is the Sagittarius stream, which is 'smoking gun' evidence of the accretion of a small companion, most probably the Sagittarius dwarf elliptical galaxy. Others include the Monoceros stream, and the so called Orphan stream. Is the stellar halo entirely formed from the accretion of satellite galaxies through the ages, or is there a genuine stellar halo, formed during the first collapse of gas in the galaxy potential?

The chemical abundances can help to answer this question. Today the dwarf satellites orbiting the Galaxy are relatively poor in alpha elements with respect to iron, while the halo stars have a high [α/Fe] (Figure 1.19). This tends to suggest the existence of a large fraction of a genuine halo. However, the answer might not be so simple, since we do not know the abundance ratios in alpha elements of the primordial dwarfs that were swallowed billions of years ago by the Galaxy.

36 Mysteries of Galaxy Formation

Also, we have been able to determine the abundances for only the stars in the solar neighborhood, which belong to the external halo, and representing only a small fraction of the total.

Soon, the astrometry satellite GAIA will be able to determine the proper motions of most stars in the Galaxy, and the whole of its 3-D dynamics will be unveiled. This will mean a great advance in our understanding of the formation and evolution of the various components of the Galaxy.

2 Infant galaxies in their cocoons

Distant galaxies were first of all investigated by studying the emission of very bright lines caused by ionized gases, known to be present in all star-forming objects. However, this was unsuccessful, because the principal (intense ultraviolet) line was completely absorbed by dust.

Now, on the other hand, astronomers make use of this absorption, which produces a characteristic 'break' in the spectra of distant galaxies (the Lyman break); its position enables us to identify the redshift and therefore the distance of the galaxies.

It is essential to know the distribution in wavelength of the energy emitted by a galaxy: two peaks stand out, one in the visible domain due to radiation from stars, and the other in the infrared, at about 100 microns, due to radiation from dust warmed by the young stars embedded within it. In a normal galaxy, these two peaks are more or less equivalent, but 'starburst' galaxies distinguish themselves by showing a much more prominent peak in the infrared.

Infant galaxies are the most likely sites for exuberant star formation, their young stars still enveloped within clouds of gas and dust. Their radiations escape only in the infrared, but the redshift effect causes the emission peak to appear in the millimeter domain. So it is microwave radiation that makes it possible to seek out these distant galaxies, hidden within their dust clouds. At these wavelengths, distant galaxies are brighter than nearer ones!

Searching for distant galaxies

The search for the earliest galaxies in the universe is compared by some to a quest for the Holy Grail. For a long time, astronomers tried to detect these objects by studying emissions in the so-called 'Lyman-α' hydrogen line. This line is emitted by atoms which, having been excited and ionized by UV radiation from young stars, revert to their fundamental state[1]. Normally, the Lyman-α line is the most intense line that we would expect to see when observing these objects, where stars should be forming in large numbers.

The first objects to collapse in on themselves, a hundred million years or so after the Big Bang, were a thousand times more massive than the Sun. They represent the 'building blocks' of the galaxies, and went on to form small galaxies as they coalesced. During these mergers, which were quite violent, bursts

38 Mysteries of Galaxy Formation

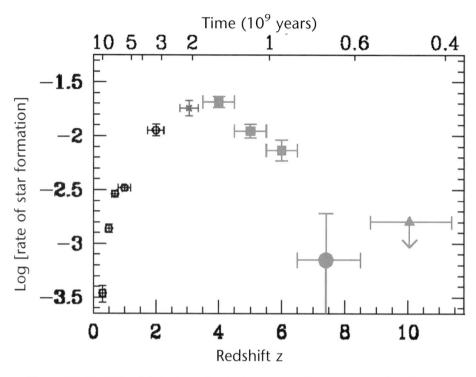

Figure 2.1 The history of star formation in the universe. The mean rate of star formation in the universe is represented as a function of time (upper horizontal axis), or of redshift z (lower horizontal axis), and is expressed in solar masses per year and per unit volume (millions of pc^3). This rate is estimated from the ultraviolet flux of deep-field galaxies as observed by the Hubble Space Telescope. Young, massive stars emit intense ultraviolet radiation, and it is possible to transform the received UV flux into an equivalent rate for the formation of stars. This interpretation also takes account of the observed colours of the galaxies. The rate of star formation must be corrected for the effect of extinction, which prevents us from seeing all star-forming regions, and especially those at high redshifts (z). The error bars are, of course, quite wide in the case of remote galaxies (after Bouwens and Illingworth, 2006).

of new stars must have occurred. As the fragments merged, and galaxies became more massive, there would have been a series of such starbursts, exciting and ionizing the gas, and reinforcing the Lyman-α emission. These proto-galaxies, or young galaxies, can be found in regions representing the first few billion years of the age of the universe, i.e. at redshifts between $z = 10$ and $z = 2$. Their associated Lyman-α line, with a laboratory wavelength in the ultraviolet of 1,200 Å (0.12

---

[1] All lines or transitions involved at the fundamental level of the hydrogen atom are known as Lyman lines, after Harvard physicist Theodore Lyman, who discovered them in 1906. The first line in the series is called Lyman-α, (then Lyman-β, etc...)

Infant galaxies in their cocoons 39

microns), becomes detectable in the visible domain, as it is redshifted towards wavelengths between 1.1 microns (near-infrared) and 0.36 microns (near-UV).

It is certain that, in their youth, galaxies saw much more star formation than occurs nowadays. So if we look at the average rate of star formation in a large galaxy such as our own – some few stars per year – it cannot explain the existence of all the stars that have by now accumulated in galaxies. Other observations have led to a general picture of star formation in the universe, and show that the mean rate of formation was at least ten times greater some nine billion years ago (at $z = 1$), and was perhaps at its maximum twelve billion years ago (at $z = 3$). The mean rate beyond $z = 3$ is still unclear; the estimate is represented in the 'Madau' diagram (Figure 2.1), named after the first astronomer (Piero Madau) to create it in 1996.

Unfortunately, searches for galaxies with strong Lyman-α emission lines met with little success early on, in spite of long periods of observation. Only very recently, thanks to the availability of more powerful instruments, have Lyman-α 'campaigns' been able to discover a whole class of remote galaxies. However, another method has had much greater success: that involving the sharp cut-off of continuous ultraviolet emission, beyond the Lyman limit, at 912 Å. This wavelength corresponds to the minimum energy a photon must have to be able to ionize the hydrogen atom.

If a distant object emits such a photon, this photon will be absorbed by any hydrogen atom along the line of sight between the remote object and the observer. But only absorbing agents of lesser redshift than the transmitter will do this; of course, the hydrogen atoms beyond (and thus of greater redshift), will not. We can therefore see that this Lyman limit can give us information about the redshift of the source. The probability of having encountered an atom of hydrogen is greater, the further away the emitting object is. This is why the method has been very successful in the case of proto-galaxies. Absorption by gas of the continuous radiation emitted by these galaxies, along the line of sight, causes a sharp decline in this radiation. This is known as the Lyman break. We therefore need only to observe the object through several filters, or in several colors, to observe whether certain objects, bright in one band, are suddenly and completely absent in the adjacent band. The position of this break reveals the redshift of the object, and therefore its distance and age.

With the filters available on the Hubble Space Telescope, it has been possible to identify a significant population of galaxies exhibiting obvious 'starburst' activity, at a redshift of $z \sim 3$. Spectroscopic examination of some of these objects has confirmed their redshifts. Incidentally, it has also shown that the Lyman-α line of the objects is not at all marked, which explains the low yield of the method described. Why is this? It is certainly because the Lyman-α line is so readily absorbed by dust, and by hydrogen gas in the surroundings (self-absorption). Unlike modern-day galaxies, which form few stars, these proto-galaxies and young galaxies had a much greater reserve of gas, and the starburst activity is always hidden within the 'nursery' clouds. In a deep-sky field, the number of very remote galaxies with redshifts greater than 3 is rather low (just a few percent of all the objects present), and the criterion of color is very useful in picking them out. This method has led to considerable advances in the study of these remote objects.

Large Lyman-α mapping projects

In fact, as techniques have progressed, and major detection campaigns have been instituted, young galaxies emitting Lyman-α lines have indeed been discovered, but they are fewer in number than was expected, being typically 100 times less intense than theoretically predicted. At least two methods are used for this research:

- long-slit spectroscopy, which involves creating the spectrum by dispersion of the light from an area of the sky delimited by a slit; but this method takes a long time, as it requires one slit per galaxy;
- wide-field narrow-band imaging, which involves making an image of an area of the sky through a color filter, centered on the frequency of the line. The filter is narrow, which means that the line is undiluted within the neighboring continuum, but no information is given on the form of the line. This technique makes it possible to cover a large area, within a limited redshift domain. The wavelength band which the filter allows to pass corresponds to the Lyman-α line for the selected redshift. The redshift is in general around $z = 4-5$, as it shifts the Lyman-α lines into the visible.

Such research has shown that the density of these objects is of the order of one per square arcminute of the sky, for redshifts between 4 and 5, i.e. nearly 100 times fewer than the galaxies discovered by the Lyman break technique. The technique has been successfully used in the case of redshifts as great as $z = 6.5$, and the detection of Lyman-α emitters at these distances makes it possible to deduce that the re-ionization of the intergalactic medium started very early, well before the epoch corresponding to redshift 6.5. Indeed, if the medium remained significantly neutral, it would absorb Lyman-α photons and remote emitters would be undetectable.

The galaxies thus discovered by narrow-band imagery have a considerable 'integrated width' (product of the intensity and velocity width of the profile). Spectroscopy shows that, in the majority of the cases, this is due to a high velocity width, rather than to any great intensity. Certainly the method used selects such objects preferentially, because narrower lines would be diluted in the band observed, and rendered undetectable. The large velocity width of these profiles might suggest emissions from accretion disks around massive black holes in active galactic nuclei. But searches for X-rays, whose emission is predicted for these very energetic sources, have had negative results, which precludes the idea that the emissions come from such nuclei. There exist today at least three hundred young candidate galaxies detected with this narrow-band Lyman-α filter technique. A few tens of these are spectroscopically confirmed, using large ground-based telescopes with apertures of around 10 m. Of course, spectroscopy consumes much observing time, as it is necessary to divide the light which comes from an object into several frequency channels (or velocities, through the Doppler effect). This spectroscopy gives some insight into the mystery of the great velocity widths of the profiles.

When spectroscopy of these objects is accurately carried out, it is seen that the profiles are very asymmetrical. This is also the case for low-redshift Lyman-α profiles. This could be due to two factors:

- the Lyman-α line is very prone to absorption by dust, which suppresses the photons by absorbing them, especially in the center of the line where the photons are 'trapped' by multiple scattering. But if the interstellar environment is very porous, certain photons can nevertheless escape at the edges of the line, and the various additional scatterings widen the profile;
- more probably, starburst activity is accompanied by stellar winds and high-velocity ejections of gas. These ejections explain the widths of lines through the Doppler effect. Moreover, the shift in velocity of the hot hydrogen gas makes it possible for the Lyman-α photons to escape and become dissociated more readily from the galaxy. Typical displacements of 300 km/s are observed between the center of the Lyman-α line and that of the stars of the galaxy. The ejection is seen asymmetrically due to the effects of absorption by dust.

In order to understand better the physics of these very early galaxies, seen at the edge of visibility, it is of interest to try to find equivalent objects closer to us. To this end we must study dwarf galaxies, poor in heavy elements or 'metals,' and therefore in dust, which show significant Lyman-α emission. Although we can observe ultraviolet Lyman-α lines in remote galaxies from the ground, thanks to their redshifts, it is necessary to venture into space to observe this line in nearby galaxies. The profiles of this line detected by UV satellites are in fact very broad, widened by gas ejections due to violent starburst formation events. Figure 2.2 shows a diagrammatic model of the scheme we can envisage by

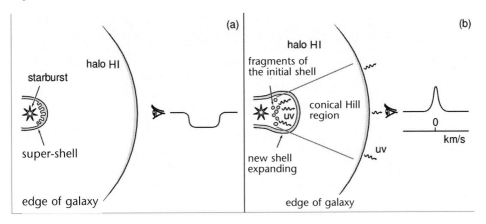

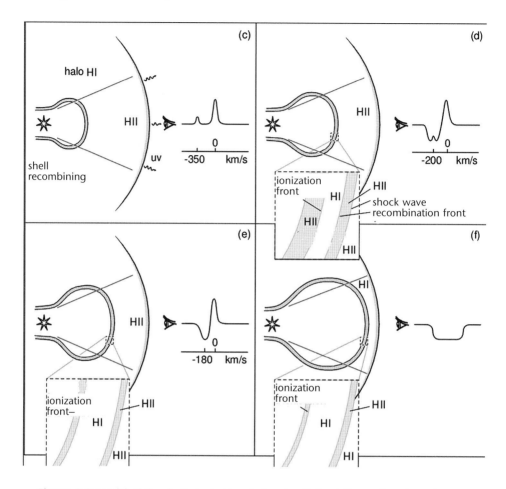

Figure 2.2 Model of the shell ejected by starburst activity at the center of a galaxy, surrounded by neutral atomic gas (HI). Panels a) to f) correspond to evolution over time, and the expansion of the ionized zone. The shape of the observed spectrum is indicated to the right of the observer's eye. First of all, stars form at the center, and ionize the surrounding gas (the ionized hydrogen, or HII, region). The halo of neutral gas, within which the starburst activity develops, absorbs Lyman-α photons and the profile obtained is that of wide absorption a). Gradually, the photons ionize the surrounding gas halo, which is diffuse, but do not encroach upon the gas disk, which is denser and more optically opaque. An observer whose line of sight does not pass through the disk will be able to see a strong Lyman-α emission line b) and c), but if the line of sight passes through the disk, the observer will see a P Cygni profile composed of one emission line and one absorption line d) and e). Recombinations within the gas swept out of the halo become ever more numerous, and the ionization front will be trapped by the recombination shell d). During the most highly evolved period, the ionized gas shell extends well beyond the galaxy's disk. There may be a double shell where the gas recombines and emits Lyman-α photons. There are several possible configurations, depending on the geometry, and the density of the neutral gas (P Cygni profile or saturated absorption as in f) (after Tenorio-Tagle et al., 1999).

observing nearby dwarf galaxies, with 'blazes' of star formation. Several distinct phases can be distinguished, from the formation of central stars to the ejection of matter in a shell around the center.

Depending on the various geometries, the Lyman-α line may or may not be visible, and may show an emission or an absorption profile, or a mixture of both. This is very similar to the profiles obtained from stars with opaque, expanding stellar winds, the prototype being the star P Cygni. The part of the spectrum in absorption is due to the region of the stellar wind between the observer and the star. This kind of profile is currently called the 'P Cygni profile.'

Energy distribution within a galaxy

How is the energy radiated by a galaxy distributed in wavelength? In order to understand and recognize the signature of star formation, let us look at the energy spectrum of typical spiral galaxies in Figure 2.3. The general shape of the curve shows two main peaks: one corresponds to radiation from stars, in the visible and the near infrared, and the other to radiation from dust heated by young stars, in the far infrared, according to the temperature reached by the dust grains. The more stars the galaxy forms, the larger is that fraction of the radiation produced by stars which is absorbed by dust. For galaxies with starburst activity, the greater part of the energy is emitted in the far infrared, at a wavelength of 100 microns. The young, bright stars are still surrounded by the cloud of gas and dust in which they were born.

Ultra-luminous galaxies were in fact first discovered during cartographical studies in the far infrared carried out by the IRAS satellite. These galaxies emit 99 percent of their energy in the far infrared, whereas in the visible domain, they seem 'normal.' After quasars, they are the most luminous galaxies in the sky, and their energy results from star formation. This phenomenon can be easily understood:

- star formation in normal galaxies today is minimal and, in the visible, we see essentially stars which have long since emerged from the parent interstellar cloud. Absorption by dust is very weak, and the infrared peak does not exceed that in the visible;
- in the case of starburst activity, the rate of formation can be 10 to 100 times greater than normal, for a limited time of the order of a hundred million years. These recently formed stars are mostly still hidden inside their cocoon of gas and dust. Their light is not seen in the visible domain, being absorbed by dust, which re-radiates energy in the far infrared, corresponding to a dust temperature of the order of 20-40 K.

In the case of proto-galaxies, the same phenomenon must intervene, and the energy from star formation is not radiated mainly in ultraviolet, visible or near-infrared wavelengths, but rather at 100 microns in the far infrared. However, this is the actual wavelength of the radiation of the galaxies; since they are remote,

44   Mysteries of Galaxy Formation

their radiation will be perceived by the Earth-based observer as considerably redshifted, as a result of the expansion of the universe. If we take redshift into account, the energy radiated by the galaxy will be detected in the submillimeter and millimeter domains, as microwaves.

The nature of dust

As we see in the energy distribution spectrum, dust plays a pivotal role in the energy budget of a galaxy. Can we deduce from our observations the nature of dust grains, or identify the various components of the dust in galaxies? And will these components maintain their properties through time, as the universe evolves?

Dust is formed from the enriched debris ejected as heavy elements by stars. It is therefore obvious that the first galaxies had less dust, relative to their quantity of gas, than the galaxies of today. The various components of the dust in the Milky Way are characterized primarily by their size, on which their temperature depends, and thus the wavelength at which they emit most energy. For a very long time, dust was mainly seen as an agent of extinction, at wavelengths around the visible and the ultraviolet. Given that its mass in the galaxy is only one percent of that of the interstellar gas, dust always seemed a secondary, troublesome ingredient, blocking the light of the stars, and relegated to the role of a tracer within the medium (e.g. of magnetic fields, density etc.), or at best a catalyst for the formation of molecules.

Actually, dust plays an active part in the cycle of star formation, and in the enrichment of the interstellar medium. Solid grains condense in the cool atmospheres of evolved stars, which eject them and recycle them into the medium. They are also formed during the explosions of massive stars (supernovae) at the end of their lives. They can, at the same time, be destroyed within the medium, if they encounter hot gas in shock waves, or ultraviolet radiation from stars, or if there are collisions between grains. But the grains can also increase in size by condensation inside molecular clouds, by accretion of a coating of ice. Grains will help to form stars, by radiating the heat of collapse, and will be destroyed in those stars, to be re-formed as the stars evolve. Then they will be ejected by stellar winds, and in supernova explosions.

The curve of extinction as a function of wavelength λ informs us of the size and composition of the grains. Overall, this involves a power law (λ^{-1}) with a spectacular hump in the region of 2175 Å, due to graphite. Currently, it is thought that grains are formed around a core of refractory material, primarily silicates and carbonaceous elements, with a coating of organic elements and ices (H_2O, CH_4, NH_3, CO_2...). To account for extinction across a wide range of wavelengths (from ultraviolet to the near infrared), we require grains of varied sizes, comparable to wavelengths. A complete range of sizes exists, from small solid particles to large molecules, in particular the polycyclic aromatic hydrocarbons (PAHs). These large molecules, comprising between twenty and a hundred atoms, and at least one 'aromatic ring' (i.e. a hexagonal ring of six carbon atoms), are very similar to those which occur in abundance in the soot and smoke derived from oil products.

Large molecules act as small dust grains

If the grains are quite large, they will be able to achieve thermal equilibrium if struck by an ultraviolet photon coming from a nearby star: the photon will be absorbed, and its energy shared among a great number of atoms. The temperature of the grain may be statistically calculated, as with any macroscopic body: in the stationary state, the ultraviolet energy it receives from the star is equal to the energy it emits in the far infrared.

In a normal galaxy, where mean interstellar radiation is the rule, this temperature is of the order of 18 K (or –255° C). The temperature goes up to 40 K (or –233° C) for starburst galaxies. On the other hand, if the grain is very small, as is the case with PAHs, the energy of just one ultraviolet photon is sufficient to excite the whole molecule to vibrate, and the 'grain' will achieve very high temperatures, much higher than that of thermal equilibrium. These very small grains will then be able to radiate as if they were at a temperature of 1000° C, which changes the wavelength completely. These grains will fluctuate between a very low temperature and a very high temperature, depending upon whether or not they absorb a photon. This phenomenon explains the great range of wavelengths emitted by dust.

The composition of the dust is suggested, or confirmed, by the spectral signatures observed in emission or absorption. It is probable that graphite or amorphous carbon or carbonaceous compounds are characterized by broad absorption at 2175 Å, and silicates produce a characteristic absorption at 10 microns. PAHs are responsible for the near-ubiquitous infrared emission lines at 3.3, 6.2, 7.7, 8.6 and 11.3 microns (Figure 2.3). There also exist a great number of diffuse absorption bands, attributed to the interstellar medium, but although they were discovered nearly a century ago, they are still not identified with any one kind of dust or molecule. Their provenance remains unknown. This model of dust makes it possible to account for not only the extinction, but also the scattering of light by grains, and its polarization, which occurs when the grains are not spherical: they then scatter the various polarizations differently, according to their alignment (by a magnetic field, for example).

Although only one-third (approximately) of the light from stars is absorbed by dust and re-radiated in the infrared in a normal present-day galaxy (like our own, for example), this is not the case for galaxies which have had, in the past, a high rate of star formation. It is known that practically all the light of an ultra-luminous galaxy is re-radiated by dust, whence the importance of knowing the abundance, nature and composition of that dust, in order to investigate the source of the light. Dust will also be essential in the earliest galaxies, although the abundance of heavy elements (and thus of dust) increases with time. But for

46 Mysteries of Galaxy Formation

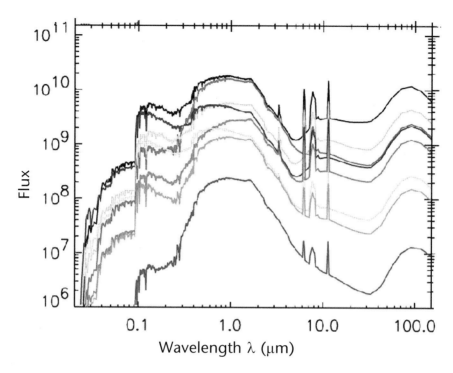

Figure 2.3 The distribution of energy in the spectra of typical spiral galaxies. Two major peaks may be distinguished: one in visible light (wavelength $\lambda = 0.5\text{-}1$ micron) corresponding to the maximum energy radiated by stars, and the other corresponding to energy in the far infrared (100 microns), corresponding to radiation from dust heated by stars. This second peak is very weak in galaxies where star formation is almost quiescent, since the radiation comes essentially from old stars which have long since left the interstellar cloud in which they were born, and are therefore not affected by extinction. However, the dust peak is by far the dominant one in the case of ultra-luminous galaxies, with starburst formation, where most of the energy comes from the young stars, still enveloped within their cocoons. To the left, the stellar spectra show their characteristic absorption lines. To the right, the spectrum is essentially that of dust, and the emission lines are the characteristic ones of PAHs (polycyclic aromatic hydrocarbons).

the most remote galaxies that we can observe, the abundance in 'metals' already seems quite sufficient: almost at the solar level. Does the nature of dust vary with time? According to the results of recent observations in the infrared by the Spitzer Space Telescope, emission ratios in the various bands show that the characteristic lines of PAHs are indeed present in the near infrared, at redshifts to $z = 3$. The sparse data that we possess indicate that variations with time are minimal.

More or less dusty galaxies

The properties of dust are relatively universal; however, certain signatures, such as that of absorption at 2175 Å, or PAH emission lines, disappear in certain environments. In galaxies with weak metallicity, such as our nearest neighbors the Magellanic Clouds, broad absorption at 2175 Å disappears completely from the extinction curve. This is interpreted as the result of the disappearance of carbon grains with weak metallicity. However, in this environment, the emission lines of large molecules (or PAHs) remain. Certain small dust grains and large polycyclic aromatic molecules can also be easily destroyed in hostile environments, such as areas very close to the active nuclei at the centers of galaxies, where hard radiation (ultraviolet, X-rays, gamma rays) is very intense. The presence or absence of these grains is therefore an indicator of the existence of the active core, itself often hidden by absorbing dust.

A means of detecting remote galaxies: millimeter waves

As we have just seen, starburst galaxies radiate most of their energy in the far infrared, at around 100 microns. For increasingly remote galaxies, this peak in the radiation will gradually be shifted towards submillimeter and then millimeter wavelengths. It will therefore be the preferred domain for the detection of the earliest galaxies. Figure 2.4 shows how the spectrum of a typical galaxy will be received at a terrestrial observatory, if the galaxy is receding progressively at redshifts between $z = 0.1$ and $z = 10$. In this diagram, where the scales are logarithmic on both axes, we notice that, with increasing distance, the spectrum of the galaxies is displaced both vertically and horizontally, at least until $z = 5$. This simple distance relationship is, firstly, a result firstly of the logarithmic scales, which transform the multiplicative factors into displacements. Also, the flux from remote galaxies varies as a power law $(1 + z)$, this factor also characterizing the variation of the received frequency, the redshift due to the expansion of the universe.

Figure 2.4 reveals a striking phenomenon: an inversion of the order of the various curves, in the neighborhood of a wavelength of 1 mm. The shift towards the right, towards the longest wavelengths, compensates for the decrease in the flux received as the galaxies recede. Although the intensity of the sources is much weakened with distance (at large values of z), the slope of the spectrum of radiation is so steep that it compensates for the fall in luminosity of the sources due to redshift. At millimeter wavelengths, all the sources have about the same apparent luminosity, whatever their distance, because their maximum emission progressively enters the domain. This phenomenon is called 'negative K-correction,' and makes it possible to detect a great number of high-redshift galaxies at millimeter wavelengths. The phenomenon will be fully exploited by the new Atacama Large Millimeter/submillimeter Array (ALMA), which will begin its operations in 2010 on the Atacama plateau in Chile.

48 Mysteries of Galaxy Formation

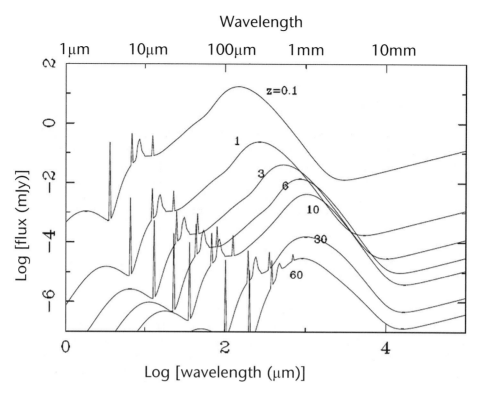

Figure 2.4 Redshift in the spectra of starburst galaxies. The energy distribution of a star-forming galaxy as received at an observatory on Earth is here schematized for ever more distant galaxies, at redshifts from z = 0.1 (upper) to z = 60 (lower). The maximum far-infrared emission due to radiation from heated dust is progressively shifted towards the submillimeter domain and then the millimeter. Lines in the mid-infrared, at wavelengths of around 10 microns, are those of PAHs (polycyclic aromatic hydrocarbons).

Figure 2.4 also shows that the various curves can be deduced by their similarity only until a redshift z = 5; thereafter they become deformed. This is due to the importance at this point of the cosmic background radiation. This background radiation, a vestige of the Big Bang, is a black body whose temperature T decreases with the expansion, as T = 2.73 (1 + z). Today (z = 0), this temperature stands at 2.73 K, much lower than that of dust heated by stars. But at z = 5, this temperature is 16 K, and becomes comparable to the temperature of some of the dust.

When we observe radiation from an object in the sky, it is observed by comparison with the sky background: we always subtract the underlying cosmic radiation background. The radiation from the object is thus the surplus emission compared with the black-body radiation bathing the whole universe. This surplus can be calculated, by supposing for example a number of stars heating the dust. At high redshifts, the temperature of the dust in distant galaxies will be higher, because it has been preheated by the black body. The resultant spectra of remote galaxies are therefore 'deformed' compared to the spectra of present-day galaxies.

The results of millimeter wave research

Deep cartographical surveys have been carried out in several areas of the sky, using bolometers at wavelengths between 0.5 and 1 mm, with a view to detecting continuous thermal radiation from dust in high-redshift galaxies. The areas in question were chosen for their absence of nearby objects, in order better to detect remote objects; hence the use of the term 'empty fields.' Today's telescopes are not large enough to carry out this research. Their diffraction limit is about 15 seconds of arc, which leads to problems in the identification of 'overlapping' sources. Hundreds of sources have been discovered, and some of them have been identified with galaxies already detected in the visible domain, of known redshift and therefore of known distance. The number of sources is about one per square arcminute.

Surveys in the visible domain find approximately 100 times more of these remote sources, by the 'Lyman break' technique (see above). But the great advantage of working in the millimeter domain, in spite of the current lack of sensitivity, is that there is no trouble with absorption by dust, and thus it does not offer a distorted view of the evolution of the galaxies. In order to increase sensitivity, some research has been carried out in areas behind foreground galaxy clusters, with a view to amplifying the radiation of remote sources using the 'gravitational lensing' effect. As shown in Figure 2.5, the light of the foreground cluster galaxies does not obstruct the detection of the remote galaxies at all, because they remain invisible in the millimeter domain.

Once the objects have been identified at other wavelengths, e.g. radio, visible or infrared, it is possible to look for emission lines due to molecular gas, in these young galaxies which are all 'starburst' candidates, or quasars, or both. At the current level of sensitivity, it is not possible to detect 'quiet' galaxies; only ultra-luminous galaxies can be caught in the net! The nature of the radiation sources will be better known when additional information is available.

Either: the dust that has been detected is heated by stars, which are formed at an extremely rapid rate, in a very compact area. This is the case with ultra-luminous galaxies, radiating 100 to 1000 times more luminous energy than the Milky Way.

Or: the dust detected is heated by an active nucleus, a quasar drawing energy from the accretion of mass by a black hole, an even more powerful source.

In the majority of cases, both phenomena are present, i.e. 'starburst' formation accompanies the 'feeding' of the nucleus and the associated luminous phenomena. The question is: in what proportion do the two emission mechanisms co-exist? The quantity of molecular gas will make it possible to test the intensity of star formation, in order to answer this question.

50 **Mysteries of Galaxy Formation**

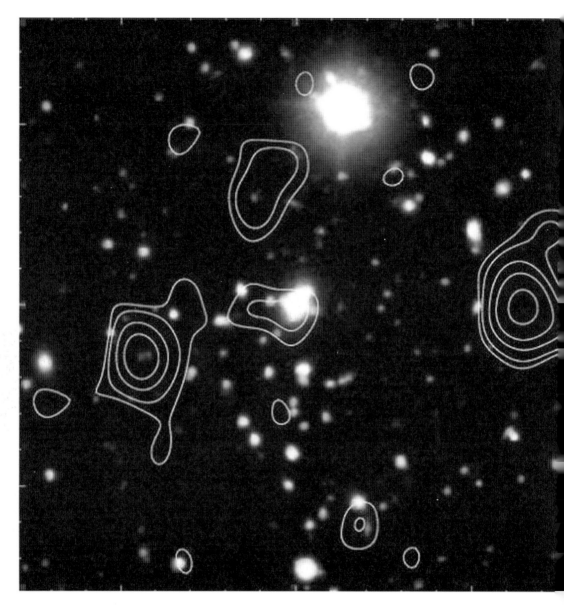

Figure 2.5 Deep-sky survey in the millimeter domain of an area behind a nearby cluster of galaxies. The nearby cluster of galaxies seen in this photo is being used as a gravitational lens, a kind of 'auxiliary telescope.' The sources discovered in the millimeter domain, corresponding to the white isophotes, have nothing to do with the nearby galaxies visible in the optical domain (after Ivison et al., 2000).

Infant galaxies in their cocoons

The principal constituent of the molecular gas is hydrogen. But molecular hydrogen, at this temperature, does not radiate: the symmetry of the molecule prevents it from having an electric dipole. The principal tracer used is carbon monoxide (CO), which is ten thousand times less abundant, and its series of rotational lines in the millimeter and submillimeter domains. These lines, regularly spaced in frequency, make it possible to examine sources at practically all the expected redshifts. The more remote the source is, the higher the level of excitation of the required CO line (i.e. of higher frequency); fortunately the flux emitted in these lines increases with the level of excitation, at least at the first levels. This advantageous cosmological correction makes it possible partly to compensate for the weakening of the flux with increasing distance from the source. This correction does not go as far as to make the remote objects brighter than nearby ones, as with continuous radiation, but it has nevertheless allowed modern instruments to detect CO lines in a few tens of high-redshift galaxies.

It is certainly not surprising that the great majority of these sources also detected in molecular lines are objects strongly amplified by gravitational lensing. One striking example is the 'four-leaf clover,' a quasar for which a foreground galaxy, in the same line of sight, plays the role of an amplifying lens, and causes us to see four images, as shown in Figure 2.6.

So the detection of molecular lines allows us to obtain more details on the physics of the high-redshift galaxy:

- its velocity width will give the total mass of the object, the quantity of interstellar gas, and the rate of star formation;
- the comparative masses of the bulge of the galaxy and the mass of the black hole as deduced from the luminosity of the quasar (assuming that it emits at the maximum Eddington luminosity) will allow us to determine the respective mass accretion rates of galaxies and supermassive black holes.

In 1995, the detection of high-redshift, ultra-luminous objects containing considerable quantities of interstellar gas came as a surprise. Nobody had expected to find evolved objects at such an early stage in the history of the universe, especially since their detection suggested the presence of a certain metallicity, in objects which had not harbored the many generations of stars needed to enrich them in heavy elements, such as carbon and oxygen (C and O). The detection of CO molecules shows that (at least) ultra-luminous objects, with 50 times more gas than in the Milky Way, were able to form very quickly, in the first billion years after the Big Bang. Certain quasars have even been detected by CO lines at a redshift of $z = 6.4$. This is very promising in the context of the new millimeter/submillimeter interferometer ALMA (see Chapter 6).

The perturbed morphology of the first galaxies

Thanks to the Hubble Space Telescope, and its high-spatial-resolution images, such as the Hubble Ultra Deep Field shown in Figure 2.7, it has been possible to

52 Mysteries of Galaxy Formation

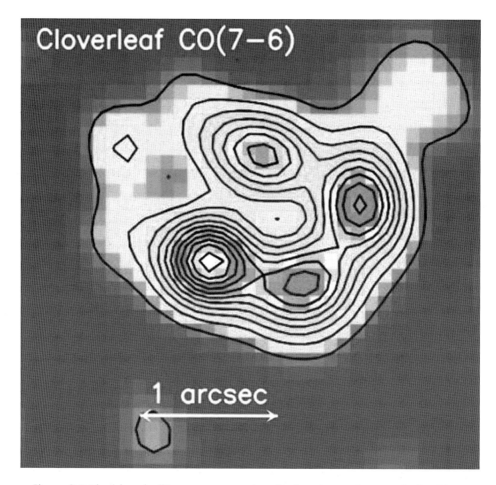

Figure 2.6 The 'cloverleaf' is a quasar reproduced as four images by a gravitational lens along the line of sight between the quasar and the observer. The contours and colours correspond to the emission in the rotational line CO (7-6) of carbon monoxide. From observations made with the IRAM interferometer on the Plateau de Bure (near Grenoble). The quasar is at redshift $z = 2.6$.

form a precise idea of the morphology of galaxies with a large look-back time in the universe.

One considerable surprise is the large number of apparently perturbed objects, difficult to classify. In the local universe, galaxies are classified along a sequence, called the Hubble 'tuning fork': it consists of one branch of elliptical galaxies, having an ellipsoidal shape, made of old stars, and opens into two branches of spiral galaxies, one for regular galaxies and one for barred galaxies (Figure 2.8). Along these two branches, galaxies are less and less concentrated in mass and luminosity. This is quantified by the ratio between the bulge and the disk, which is decreasing. At the end of the sequence are classified the 'irregulars,' galaxies

Infant galaxies in their cocoons 53

Figure 2.7 This million-second-long exposure, called the Hubble Ultra Deep Field (HUDF), reveals the first galaxies to emerge from the so-called 'dark ages,' the time shortly after the Big Bang when the first stars reheated the cold, dark universe. This view is actually two separate images taken by Hubble's Advanced Camera for Surveys (ACS) and the Near Infrared Camera and Multi-object Spectrometer (NICMOS). (NASA, ESA, S. Beckwith (STScI) and the HUDF Team). See also PLATE 5 in the color section.

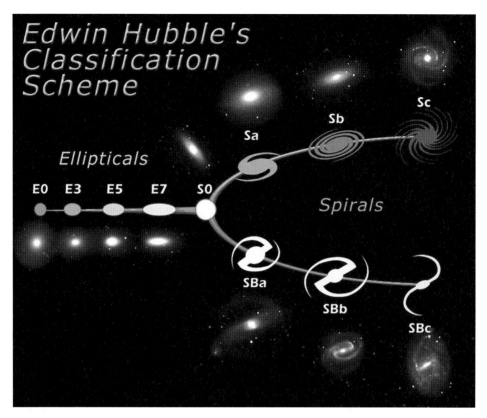

Figure 2.8 Hubble Tuning Fork diagram. Edwin Hubble was the first person to classify galaxies. Astronomers use his system, called the Hubble Tuning Fork, even today. First, Hubble divided the galaxies into two general categories: elliptical and spiral galaxies. Elliptical galaxies are shaped like ellipses, and spiral galaxies are shaped like spirals, with arms winding in to a bright center. Elliptical galaxies are classified by how round or elongated they look. An E0 galaxy is very round and an E7 galaxy is very elongated. In detail, the number after the 'E' is determined by the galaxy's ellipticity – the ratio of the ellipse's major axis to minor axis. Galaxies with higher ellipticities have higher numbers. Hubble noticed that some spiral galaxies have a bright line, or bar, running through them. He called these galaxies 'barred spiral galaxies.' Galaxies with spiral arms, but without the bar, are just called 'spiral galaxies.' Spiral galaxies are further classified by how tightly their arms are wound. Type a galaxies have their arms wound very tightly and have large central bulges. Type c galaxies have very their arms would loosely and have small central bulges.

Some galaxies are a transition type between the elliptical and spiral galaxies, labeled S0 on the tuning fork. These are called 'lenticular galaxies.' Lenticular galaxies have a central bulge and a disk but no spiral arms. The third class of galaxies is irregular galaxies. Irregular galaxies are neither spiral nor elliptical, and can have any number of shapes. They are frequently the product of two galaxies colliding with each other, or at least affecting each other through the force of gravity (NASA, STScI).

lacking a clear and coherent pattern in their morphology. They are in general of lower luminosity than the rest of the sequence, and are called dwarf galaxies.

Looking back in time, the fraction of irregular galaxies is higher and higher, and irregular morphology appears not to be confined to the dwarfs, but also to giant galaxies. Of course, some of these perturbed galaxies could be just interacting galaxies. There are good arguments to suggest that the frequency of galaxy interactions was higher in the past. First, due to expansion, the universe was denser in the beginning, and encounters were more frequent; second, and more importantly, at the epoch of half the life-time of the universe, galaxies formed groups, inside which they interacted and merged, and those groups then gathered in galaxy clusters. Observations confirm that galaxy binaries were more frequent in the past and mergers too. However, this alone cannot explain the large fraction of perturbed morphologies, and another phenomenon must be invoked. Many galaxies look like an ensemble of clumps, and do not show any spiral structure. Galaxies are increasingly clumpy when we look back in time. Their luminosity is more and more dominated by a few star complexes, one-tenth the galaxy diameter in size, and of mass equivalent to a hundred million solar masses. When the clumps reach 30 or 50 percent of their luminosity, they are called 'clumpy galaxies.' Sometimes the clumps are aligned together, and the object is called a 'chain galaxy.' Rounder systems are called clump-clusters. Certainly, these varieties correspond to different orientations along the line of sight (Figure 2.9).

The existence of clumpy galaxies is not a surprise from the theoretical point of view. Indeed, gravitational systems which are very rich in gas are violently unstable under self-gravitational forces. Young galaxies begin their lives through gas accretion, and, because of gas dissipation and cloud collisions, the gas component settles rapidly into a disk, perpendicular to the angular momentum acquired through tidal interactions between the forming structures. At this time, there is not yet any bulge. Bulges are stellar structures that gather mass, either from secular evolution of disks along time, or through multiple accretion of small companions. The first galaxy disks are therefore very unstable, against gravitational collapse. Only spheroidal stellar systems, like bulges, not dominated by rotation, would have a stabilizing influence on the disk. Also the dark matter halo has not enough gravitational influence in the center, which is dominated by the gas mass. Simulations predict the formation of massive clumps. However, the lifetime of clumps should be no more than half a billion years, in an isolated system. The high observed frequency of clumpy objects means that galaxies are still abundantly fuelled by intergalactic gas, to maintain the instability of primordial disks.

Numerical simulations reveal how clumpy the first gas-rich galaxies can be (Figure 2.10). The clumps are so big that they perturb the surrounding medium, including the dark matter, and the corresponding deformations have a braking effect on their motions. The clumps are slowed down, and they spiral towards the center, progressively to form a bulge. Clumpy galaxies can therefore be the precursors of normal spiral galaxies with bulges. This strong instability of young and gaseous galaxies explains the presence of irregular morphologies, even without any merger.

56 Mysteries of Galaxy Formation

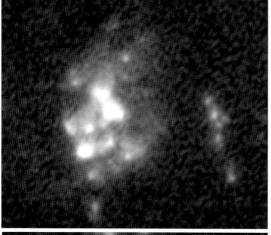

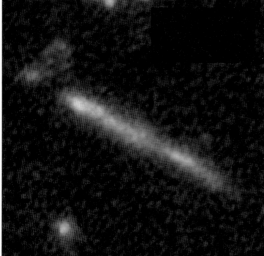

Figure 2.9 Clumpy galaxies. Galaxies become increasingly clumpy as we look back in time. The spiral and regular structure disappears, and instead the luminosity becomes dominated by a few stellar clumps, randomly distributed across the disk. When the latter is seen edge-on **(bottom panel)**, the clumps look like a chain, and the object is called a 'chain galaxy.' Rounder objects, seen more 'face-on' **(top panel)**, are called 'clump-cluster galaxies' (from Hubble Ultra Deep Field observations, courtesy NASA, ESA, S. Beckwith (STScI) and the HUDF Team).

The beginning of the story . . .

Has the frantic search for galaxies at ever higher redshifts (particularly the deep observations by the Hubble Space Telescope) revealed to us the very first galaxies to form in the universe? The existence of hundreds of objects at redshifts between $z \sim 3$ and $z \sim 4$ has been confirmed, and there are a few additional objects at distances greater than $z = 5$-6. The scarcity of candidate objects at $z \sim 6$-7 shows both the detection limits of current instruments and also undoubtedly the intrinsic rarity of remote and luminous youthful galaxies, less than a billion years after the Big Bang (Figure 2.11).

Infant galaxies in their cocoons 57

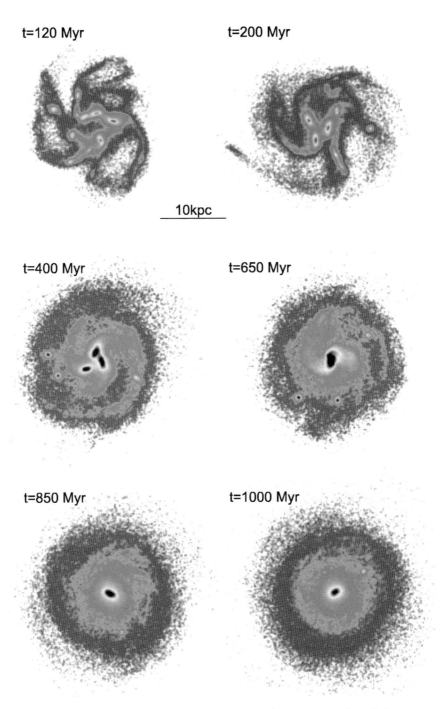

Figure 2.10 Simulations of clumpy galaxies. The first galaxies are constituted almost entirely of gas that settles in a disk, through dissipation. The disk is dynamically cold, and highly unstable against gravitational collapse into clumps. Progressively, clumps are driven into the center by dynamical friction, and form a bulge, which stabilizes the disk (from Bournaud et al., 2007). See also PLATE 6 in the color section.

58 Mysteries of Galaxy Formation

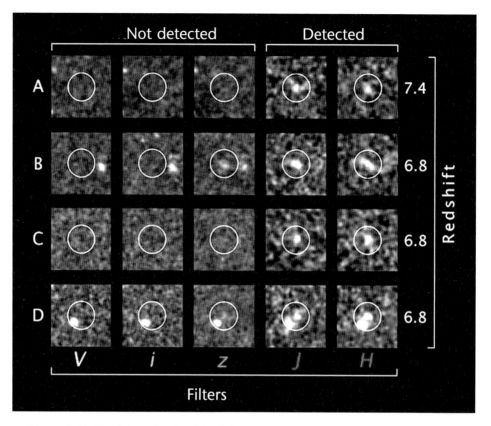

Figure 2.11 Candidates for the title of the most remote galaxies currently known. These objects have been selected for their colors in the optical and near-infrared domains, and their z-dropout frequencies. Note that the object at the center of the various images has been detected using J and H filters, but is undetected through higher-frequency filters. In order to confirm these candidates, it will be necessary to take spectra (and therefore secure a spectroscopic redshift, as well as the indicative photometric redshift), but this will have to wait until larger and more sensitive telescopes become available. Each image represents a square of 3.5 arcseconds (after Bouwens and Illingworth, 2006).

The hundreds of high-redshift galaxies now detected give some insight into the major pathways of the evolution of the universe. In the galaxies we observe, star formation proceeded at faster rates in the past than nowadays, and it has been established that most of the stars in nearby galaxies were born about halfway through the current lifetime of the universe, i.e. about 7 billion years ago. Galaxies where large numbers of stars form are generally very rich in gas and dust, such that most of the light from the stars does not escape in the visible domain, but as far-infrared radiation: the thermal radiation of dust warmed by stars. Here, then, is the phenomenon of ultra-luminous galaxies that radiate more than 90 percent of their energy in the infrared (ULIRGs). In these galaxies,

Infant galaxies in their cocoons

it is always difficult to separate the energy due to the active nucleus from the energy from star formation, but observation in hard X-rays, which comes only from active nuclei, points us to the most luminous nuclei, e.g. quasars.

Although ultra-luminous galaxies (100 times brighter than the Milky Way) are rare today, and always correspond to mergers of galaxies, they were much more common during the early stage of the universe. Progressively, bright infrared galaxies (10 times brighter than the Milky Way – the LIRG) took up the baton and began to dominate as far as star formation was concerned, and today so-called 'normal' galaxies, with only modest star formation rates, dominate. Since the universe was half its present age, there has been a steady decline in the mean rate of star formation, and this is apparent in Figure 2.1.

The techniques used to detect these deep-field galaxies have revealed their gregarious early tendencies: even at $z = 3$ (just two billion years after the Big Bang), they have already formed into groups and clusters, at a much earlier epoch than might have been expected from models of the formation of structures. In fact, these traditional models predict that small structures form first, and then come together to form larger ones such as galaxy groups and small clusters, which today have become superclusters, etc. These models doubtless concentrate essentially upon the mass of structures, which are dominated by dark matter, while the behavior of visible (or 'baryonic') matter is more difficult to predict. If we take into account the dissipation of gas, its gravitational collapse to form stars, and self-regulatory phenomena (stellar winds, supernovae, etc.), luminous phenomena are very complex and their distribution can sometime appear to be in apparent contradiction with underlying matter.

Several kinds of galaxies have been discovered and, as occurs in all new (and not thoroughly understood) fields of study, a well-stocked nomenclature has arisen, but the various categories of objects have not yet settled into a common theoretical scheme offering a unified approach to our understanding of their origin. Because of redshift, detection techniques tend to concentrate on long wavelengths, but one category of extremely red galaxies has often been observed: the EROs (Extremely Red Objects). It seems that these galaxies are a mixture of young galaxies forming many stars reddened by vast quantities of dust, and very old stellar systems, with a great mass of stars in a very ancient population.

One considerable surprise that has emerged from these observations has been the discovery of very massive systems that evolved very early on in the universe. According to the theory of the hierarchical formation of galaxies, massive galaxies are expected to occur only at a later stage, as smaller galaxies merge; but it does seem that, on the contrary, the first galaxies to form were the most massive, and thereafter they evolved very 'passively,' with no new star formation, until the present day. These are the elliptical galaxies, predominantly red, which appear to have no gas left and are therefore more or less 'dead.' The discovery of relatively 'old' galaxies existing so early on in the universe shows us that evolution can proceed very rapidly in certain environments.

These discoveries have led some astronomers to question the idea of the hierarchical merging of structures, and they prefer that of the very rapid 'monolithic' collapse of the first elliptical galaxies. The debate is far from concluded, since star formation must have begun within galaxies of modest dimensions, redshifted to invisibility for our modern instruments; these galaxies then merged to form massive ellipticals. The characteristic time necessary for the interaction and merging of structures was much shorter at the beginning of the universe.

Questions still unanswered

There are two fundamental questions to ask about the formation of galaxies:

- When were the stars formed?
- When was the mass of galaxies assembled, and when did the galaxies acquire their present morphology?

One part of the answer might come from the morphology of the galaxies. Now, the morphology of high-redshift galaxies is not easy to reconcile with the usual classification according to the Hubble sequence. Certainly, we see many disks and bulges, and therefore many spirals and ellipticals, but the number of 'irregular' galaxies is considerably increased. Deep-field images obtained by the Hubble Space Telescope (Figure 2.12) clearly show perturbed morphologies, like those of interacting galaxies nearer to us. Certainly, we must be cautious in our interpretation of these results, since the effects of redshift oblige us to observe remote galaxies principally in the ultraviolet, which corresponds to star-forming regions, and these are often irregularly scattered within the galaxies. However, these morphologies occur in the great majority of near-infrared images.

Statistically, it has been possible to quantify the fraction of interacting and merging galaxies as a function of redshift. There are several methods, involving for example the enumeration of galaxy pairs, or the quantifying of the asymmetry and perturbations seen on images of galaxies.

The further back in time we look, the more spectacular the increase in interactions, as revealed in the diagram in Figure 2.13. Between $z = 0$ and $z = 1$, the rate of interactions increases by a factor of nearly 100, and is maintained as far out as $z = 3$. A typical galaxy such as the Milky Way will thus have experienced four or five major galaxy mergers since the epoch when the universe had attained only 20 percent of its current age: its mass may well have been increased by a factor of 10. However, there are many variations involving the mass and the environments of galaxies, and there remain great uncertainties about their values.

Another class of objects has also been discovered by chance, thanks to the narrow-band Lyman-α filter technique described at the beginning of this chapter. These objects are the LABs (Lyman-Alpha Blobs), huge concentrations of neutral atomic hydrogen emitting the Lyman-α recombination line. The surprise is the very size of these objects, which are among the largest known in the universe: gaseous structures often 300 million light-years across! Such a size corresponds to that of enormous galaxy clusters, or even superclusters. Yet they are detected at

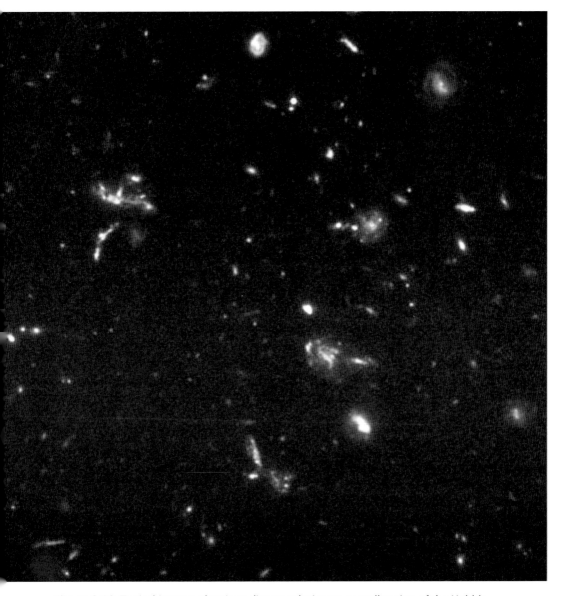

Figure 2.12 Typical images of various distant galaxies on a small region of the Hubble Ultra Deep Field (HUDF) image. Note the very irregular morphologies of these galaxies. (NASA, ESA, S. Beckwith (STScI) and the HUDF Team).

redshifts greater than 3 (representing a time only two billion years after the Big Bang). These collections of gas clouds are aligned as if on cosmic filaments, in clumps the size of galaxies or groups of galaxies. The gas may have been ejected during starburst events, by stellar 'superwinds' on a galactic scale (Figure 2.14).

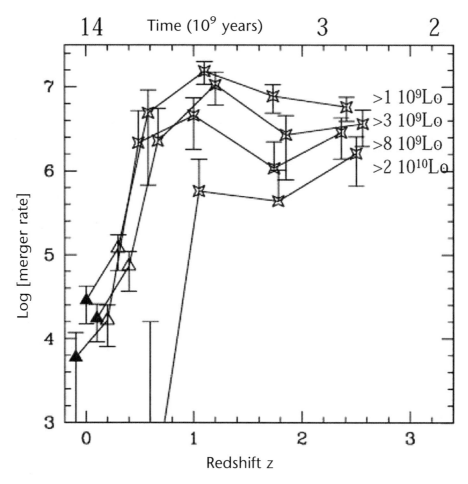

Figure 2.13 Evolution of galaxy merger rates, by billions of years and by unit of comoving volume (in billions of pc^3), as a function of redshift z, and for several categories of luminosity, indicated at top right in terms of solar luminosity (L). The rise in the number of mergers between z = 0 and z = 1 is spectacular (after Conselice, 2006).

There is another way to chart the evolution of structures, using techniques involving absorption by distant quasars. This is a very sensitive technique, employed for decades now, allowing us to sample intergalactic gas at various redshifts along the line of sight. The most intense line involved is that of neutral atomic hydrogen, the Lyman-α line, and the sensitivity is such that we can perceive a 'forest' of around a hundred lines in the direction of certain quasars. This technique has led to the quantification of the density of the gas between galaxies, and to the study of its decrease as a function of redshift. Intergalactic clouds are very inhomogeneous, and the observation of associated 'metal' lines

Infant galaxies in their cocoons 63

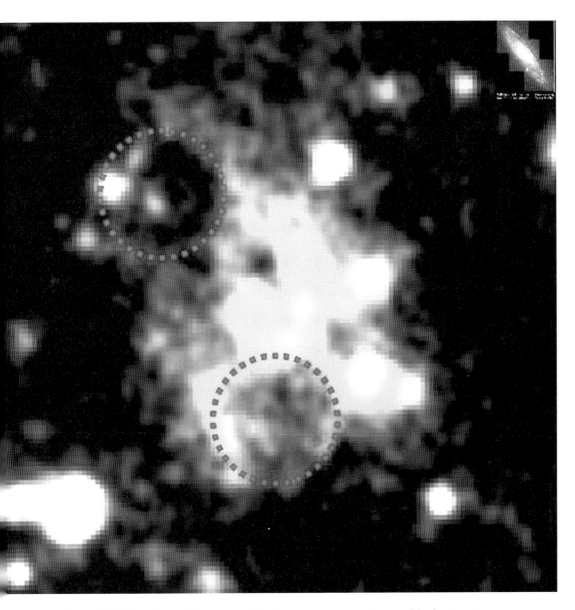

Figure 2.14 Emission in the Lyman-α line from a large area of neutral hydrogen gas surrounding several galaxies. At top right, to give an idea of size of this object, is the Andromeda Galaxy. Note the bubbles where emission is either absent or weaker, suggesting the existence of bubbles of star formation (areas within circular dotted lines) (after Matsuda et al., 2004).

64 Mysteries of Galaxy Formation

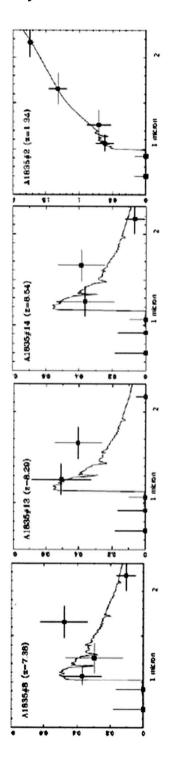

Figure 2.15 Searching for high-redshift candidates (at z > 7), with the 'gravitational telescope.' Three of the candidate galaxies, behind the cluster Abell 1835, at possible redshifts of z = 7.38, 8.29 and 8.54, are detected using 3-4 near-infrared filters, but none further away because of the high-frequency 'break' (Lyman limit). The observations (dots) are superimposed on models of spectra, enabling us to propose a redshift. On the right, a low-redshift object at z = 1.34 is shown for the purpose of comparison (after Schaerer et al., 2005).

(e.g. iron, carbon, magnesium) helps us to assess the amount of enrichment by heavy elements, which come from the stars that have formed within the galaxies.

Finally: although the sensitivity of our current instruments may not be up to the task, the search for the earliest objects continues, using a 'gravitational telescope.' This involves taking advantage of the amplification provided by 'gravitational lenses' in the line of sight. A cluster of galaxies, situated between a more distant cluster and the observer, can amplify light, in certain favorable places by an order of magnitude, thereby facilitating detection.

This is how several candidate objects were discovered behind the cluster Abell 1835, as shown in Figure 2.15. However, spectroscopy using the 8.2-m unit telescopes of the VLT (Very Large Telescope) is still pushing the instruments to their limit, and the new-generation ELT (Extremely Large Telescope) will be welcomed!

3 The origins of black holes

There are several types of black hole; notable among them are the 'supermassive' black holes.

For some years now, it has been known that every galaxy harbors a supermassive black hole at its center, its mass proportional to that of the galaxy's bulge. This suggests that supermassive black holes and galactic bulges have grown simultaneously.

The supermassive black hole grows very early on in the lifetime of a galaxy, in parallel to the history of its star formation.

In the hierarchical scenario, where galaxies form by merging, black holes are also caused to merge, and 'binary black holes' should therefore be observable.

There is no doubt that intermediate-mass black holes also exist, but it is difficult to identify them.

What is a black hole?

The term 'black hole' refers to an object that is so condensed, so compact, that its escape velocity (the velocity it is necessary to attain in order to escape from the object) is greater in its immediate vicinity than the velocity of light; therefore, even photons cannot break free from it. Such objects had been mooted as long ago as the eighteenth century, firstly in 1783 by the British geologist and astronomer, the Reverend John Michell, and thirteen years later by the great French mathematician, astronomer and physicist, Pierre-Simon de Laplace, but the development of the theory of black holes had to wait until the twentieth century. Gravity near black holes is so strong that it falls within the context of general relativity, though it is sometimes possible to deal with its values using the more familiar Newtonian formulae.

The square of the escape velocity around an object of mass M is proportional to its mass divided by its diameter R. In the case of ordinary celestial bodies, this velocity is far less than that of the velocity of light (c).

In the case of a black hole, the body has collapsed in upon itself and its radius can no longer be defined. However, its horizon R, the distance from the center at which the light disappears, the point of no return, is precisely where the escape velocity becomes equal to c. The value of this horizon R varies therefore in proportion to the mass M of the black hole[1].

It is thus possible to deduce that the compact nature, i.e. the mean density within the horizon of the black hole, is a function of the total mass of the black hole, varying as M/R^3, or as $1/M^2$. In other words, small black holes are the most compact. To give an idea of what this entails, the mean density for a black hole of a million solar masses is about 20,000 times that of water. However, for a black hole of one billion solar masses the mean density is only 2 percent that of the density of water.

Two types of black hole are observed:

- black holes of the stellar type, formed as part of the evolution of massive stars. Their mass is of the order of a few solar masses and above, and their mean density exceeds that of all known states of matter;
- black holes of the galactic type, which are much more massive. Their mean density is of the order of that of water, and may be even less, which is disconcerting. It would be possible to cross their horizons, i.e. past the point of no return, without being aware of the fact and without having felt the effects of tidal forces. The phenomenon is indeed a daunting one for any astronaut who might unwittingly travel too close to one of these black holes, but it is also has implications for those studying the luminosity of black holes at the centers of galaxies.

Black holes, by definition, ought not to be visible, but they can be among the most luminous objects in the universe. This is due to infalling matter, 'swallowed' by the black hole. At the expense of gravitational energy, matter is accelerated and radiates at a far greater level of efficiency than any other mechanism of radiation. While nuclear reactions within stars extract, at most, one percent or less of the energy of mass from matter, gravitational attraction and accretion of matter by a black hole extract around 10 percent. The matter that falls into a black hole remains for some time within an 'accretion disk,' where it progressively loses angular momentum. It is from this disk that energy is radiated away, causing the black hole to appear as a 'quasar' ('quasi-stellar object/source,' as described below) associated with remote galaxies.

---

[1] In Newtonian mechanics, the most commonly used convention for the value of the radius R of a black hole, the radius of the event horizon, is given by:

$$R = 2GM/c^2$$

Hence, the size of a black hole, as determined by the radius of this event horizon (or Schwarzschild radius) is proportional to the mass M of the black hole as:

$$R = 2.954 \, M / M_\odot \text{ km,}$$

where R is the Schwarzschild radius (in kilometers) and $M_\odot$ is the mass of the Sun. A black hole's size and mass are thus simply related. In general relativity, R is not so straightforward to define due to the curved nature of space-time and the choice of different coordinates.

The formation of stellar-type black holes was predicted not long after the nature of stellar evolution was understood (at the beginning of the twentieth century). Depending upon the initial mass of the star (which can be from 10 to 100 solar masses), and especially upon its residual mass after successive mass losses via solar winds and ejections of gas envelopes, and after the core of the star has ceased to convert hydrogen to helium, the residual core will end its life as:

- a white dwarf (the fate of solar-mass stars), within which the force of gravity is balanced by the 'Pauli pressure' (electron degeneracy pressure), with the electrons forming a gas composed of degenerate fermions;
- or, if the mass is greater, a neutron star, within which electrons fuse with protons to form a gas of degenerate neutrons, whose pressure balances the force of gravity;
- or finally, if even that pressure is not sufficient, and the core is of mass greater than 3.3 solar masses, the star collapses to become a black hole. The existence of these stellar-type black holes is confirmed indirectly through the observation of binary stars where one of the components has become a black hole. This component is invisible, but its mass (which is far greater than the limiting mass for white dwarfs and neutron stars) may be calculated from the period of rotation of its companion. The first system of this kind to confirm the existence of black holes, in the 1970s, was the X-ray source Cygnus X-1, about 6,000 light-years distant (Figure 3.1). We know of about twenty such systems in our own Galaxy.

Do black holes of the 'galactic' type exist?

Since the 1960s, when ultra-luminous, apparently stellar objects were discovered and labeled 'quasars' (quasi-stellar objects), it has been suspected that they derive their energy from matter falling into a supermassive black hole: the only mechanism capable of unleashing so much energy. In fact, the energy from a quasar at the heart of a galaxy may be two or three orders of magnitude greater than that radiated by the rest of the galaxy!

The existence of these supermassive black holes has been established more directly thanks to observations of the center of our own Galaxy. The Milky Way's black hole is the nearest of this class of black hole that we can observe, at a distance of only 25,000 light-years from the Sun! This proximity means that we can observe the stars surrounding it at high spatial resolution, one by one, measuring their orbits directly by studying their proper motions and radial velocities. This can be done only in the near infrared, a wavelength at which dust does not completely obscure the galactic center.

The orbits of stars very close to the nucleus, which has been identified with the radio source Sagittarius A*, are ellipses, the Keplerian orbits predicted around a massive 'point' source, such as those of the planets around the Sun. Since the orbits of these stars have been observed so close in to the nucleus, it has been

70 Mysteries of Galaxy Formation

Figure 3.1. Accretion disk around a black hole in a binary star system, such as Cygnus X-1. This artist's impression shows an 'ordinary' star (left) whose gaseous envelope is so drawn out that it has reached the Roche lobe, the boundary between the respective spheres of attraction of two stellar companions. The matter lost by the star is accreted onto its companion, which has already collapsed to become a black hole (right). This matter does not fall directly into the black hole, but revolves around it, having to lose its momentum angular (rotation) before being able to cross the black hole's horizon. During this period, the matter heats up and radiates very energetically, in wavelengths as short as X-rays and gamma-rays. This is the mechanism that enables black holes to be 'seen' (NASA/STScI and ESA). See also PLATE 7 in the color section.

possible to establish the ultra-compact nature of the central source: the density is higher than might be expected from a cluster of stars (even if they were neutron stars), or any other exotic body that theory might embrace. This central mass is equal to 3 million solar masses (Figure 3.2).

Stellar and galactic black holes are observed more or less directly. It is certain that between the stellar and galactic types a range of intermediate black holes exists, but this mass range is very difficult to identify. We shall return to this point later.

Might there be black holes less massive than the objects formed during the final stage of the lives of stars? Very small primordial black holes, mini- or micro-black holes, could have been formed during the Big Bang, but were unstable. It should be remembered that black holes have the equivalent of a 'temperature,'

The origins of black holes 71

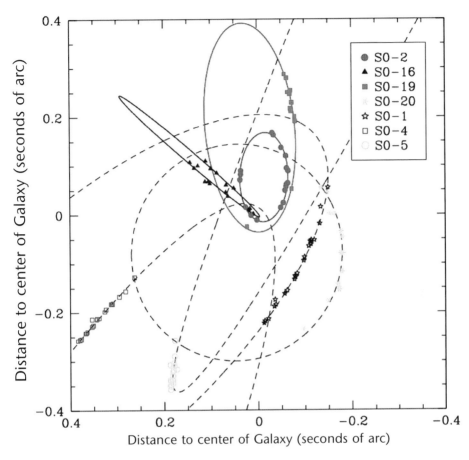

Figure 3.2. The 'ballet' of stars at the center of our Galaxy. Measurements of the proper motions of the various stars in the plane of the sky, added to measurements of velocities along the line of sight as deduced using the Doppler effect, enable us to reconstruct the stars' motions in three dimensions, and thereby learn the mass of the central object. This object is so compact that it can only be a black hole, of an estimated 3 million solar masses. On the diagram, different symbols represent observed positions of seven stars tracked over about ten years and superimposed on their calculated orbits in the vicinity of the nucleus of our Milky Way galaxy. The average velocity of these stars is 1000 km/s or greater (that of the Sun being only 200 km/s around the galactic center). The center, Sagittarius A*, corresponds to the center of the coordinates (0,0). The orbits of the stars S0-2 (filled circles) and S0-16 (triangles) were observed to approach the black hole very closely; in the case of S0-16 to a distance of less than 45 astronomical units (6.7 billion km), at a velocity of 12 000 km/s. This distance corresponds to 600 times the radius of the black hole's horizon (11 million km) (after Ghez *et al.*, 2005).

72 Mysteries of Galaxy Formation

and the smaller the black hole, the higher it is. Mini-black holes therefore evaporate very rapidly, through ejection of particles, and are unlikely to have survived to the present day.

The search for compact objects of less than one solar mass has been carried out in the context of the investigation of candidates for the missing mass around the Milky Way. This research has been based upon the use of gravitational micro-lenses: objects that, although invisible themselves, should bend rays of light passing in their vicinity. The experiment has been negative. The only gravitational micro-lenses observed have been low-mass stars.

Black holes and galaxies

Until very recently, only galaxies with active nuclei, which are in the minority among galaxies, were thought to harbor supermassive black holes at their centers. Active galactic nuclei (AGN) are of various kinds: the most energetic are the quasars, far brighter than whole galaxies. The AGN phenomenon is detected essentially though analysis of the emission spectra of galaxies.

Most galaxies exhibit emission lines due to ionized gas. Half of them owe this emission to the formation of stars, whose ultra-violet radiation is ionizing. The other half show evidence of a different kind of activity in the nucleus. Clues are present in the spectrum, which is very wide in AGN, much wider than might be expected from the galaxy's rotational velocity, and from the degree of ionization of the gas. This activity may show various degrees of intensity.

The most luminous active galactic nuclei are quasars whose energy, derived from the matter around the black hole, is 10 to 1,000 times greater than the luminosity of an entire galaxy (Figure 3.3). Today, only one percent of all galaxies are quasars. However, this percentage was much greater in the past. Another kind of galaxy is the Seyfert galaxy, an AGN showing weaker activity (Figure 3.4). About 10 percent of all galaxies are Seyfert galaxies. The remaining 40 percent are transitional objects, with a low level of activity in the nucleus and a low rate of star formation.

How many black holes are there in the universe?

The energy of AGN comes from the gas falling into the central black hole. While losing gravitational energy, the gas gains kinetic energy and, by means of shock waves, dissipates it by radiation in a very wide spectrum of wavelengths. Simultaneously, it also feeds the black hole, which grows. The black hole may also feed on stars that venture too close to it: tidal forces destroy them and they revert to gas, which is then absorbed by the accretion disk. However, once all the surrounding matter, gas or stars, has been swallowed, the now 'unfed' black hole will no longer be visible. How many black holes are there in the universe, lurking at the centers of galaxies?

The origins of black holes 73

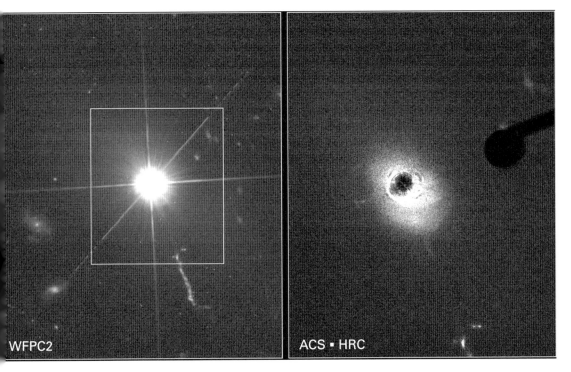

Figure 3.3 Left: this image of the quasar 3C273 in the constellation of Virgo, acquired by the Hubble Space Telescope's Wide Field Camera 2, shows the brilliant quasar but little else (NASA and J. Bahcall (IAS)). **Right:** once the blinding light from the brilliant central quasar is blocked by the coronagraph of Hubble's Advanced Camera for Surveys (ACS), the host galaxy pops into view (NASA, A. Martel (JHU), H. Ford (JHU), M. Clampin (STScI), G. Hartig (STScI), G. Illingworth (UCO/Lick Observatory), the ACS Science Team and ESA). See also PLATE 8 in the color section.

The two (extreme) possibilities that could explain the frequency of quasars and AGN are:

- *either*, black holes exist in only a minority of galaxies; they are regularly supplied with new material, and it is they that cause the phenomenon of active nuclei;
- *or*, on the contrary, black holes are very common in the universe, but are usually invisible, and they are supplied only very rarely with material, which would also account for the low frequency of observed AGN.

For a decade now, we have seen confirmation that the second hypothesis is indeed the correct one. Clues so far, involving the mass of observed black holes, have pointed the way. If it is only those galaxies possessing a supermassive black hole that are supplied with new material, then we would expect to see only a few rare and very massive black holes, of some billions of solar masses or more, for only the richest would be supplied! But the demography of quasars and AGN is

74 **Mysteries of Galaxy Formation**

Figure 3.4 Hubble Space Telescope face-on view of the small spiral galaxy NGC 7742. This is a Seyfert 2 active galaxy, a type of galaxy that is probably powered by a black hole residing in its core. The core of NGC 7742 is the large circular 'yolk' in the center of the image. The lumpy, thick ring around this core is an area of active starbirth. The ring is about 3,000 light-years from the core. Tightly wound spiral arms also are faintly visible. Surrounding the inner ring is a wispy band of material, which is probably the remains of a once very active stellar breeding ground (The Hubble Heritage Team (STScI/AURA) and NASA). See also PLATE 9 in the color section.

not like this. On the contrary, they exhibit a whole range of masses, and very massive black holes are rare.

Also, observations by the Hubble Space Telescope since the 1990s have revealed central black holes in several nearby galaxies, and even in non-active ones. High spatial resolution has meant that we can measure the velocity dispersion of stars very close to the nucleus, and thereby deduce its mass. For example, the enormous velocities observed at the center of our neighboring galaxy, the Andromeda spiral, suggest the existence of a black hole of 70 million solar masses.

When we examine statistically all the results from recent years, we can see a clear correlation between the mass of the black hole and that of the bulge of the galaxy that harbors it (Figure 3.5). This is a proportional relationship, the mass of the black hole being 0.14 percent of that of the bulge.

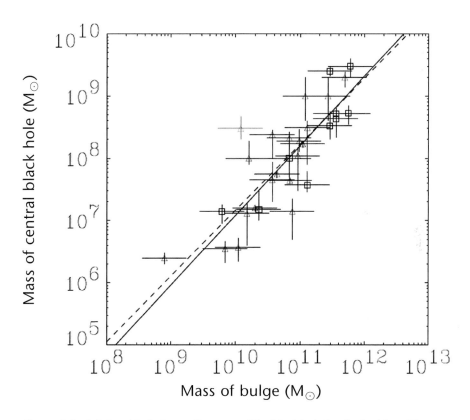

Figure 3.5 Relationship between the mass of the black hole (vertical axis) and the mass of the bulge of a galaxy (horizontal axis). The slope is close to 1, on logarithmic scales, signifying that the mass of the black hole is proportional to the mass of the bulge, with a proportionality ratio of 0.14 percent (after Haering and Rix, 2004).

76 Mysteries of Galaxy Formation

Figure 3.6 Streaming out from the center of the galaxy M87 is a black-hole-powered jet of electrons and other sub-atomic particles traveling at nearly the speed of light. At first glance, M87 (NGC 4486) appears to be an ordinary giant elliptical galaxy, one of many in the nearby Virgo Cluster of galaxies. In the 1950s, one of the brightest radio sources in the sky, Virgo A, was discovered to be associated with M87 and its jet. Lying at the center of M87 is a supermassive black hole, with a mass equivalent to over 2 billion times the mass of our Sun. The jet originates in the disk of superheated gas swirling around this black hole and is propelled and concentrated by the intense, twisted magnetic fields trapped within this plasma. The light that we see (and the radio emission) is produced by electrons twisting along magnetic field lines in the jet, a process known as synchrotron radiation (NASA and the Hubble Heritage Team (STScI/AURA)). See also PLATE 10 in the color section.

The origins of black holes

It is interesting to note that the disk of a spiral galaxy does not figure in this relationship; it is only the bulge that is involved. With elliptical galaxies, which may be seen as spheroids, or as bulges in themselves, the mass of the black hole is proportional to the total mass. This explains why the largest black holes and the largest AGN are usually associated with elliptical galaxies, such as M87 in the Virgo cluster (Figure 3.6), or with early-type spiral galaxies (showing large bulges).

Practically all galaxies have a supermassive black hole at their center. The formation of these black holes is therefore an integral part of the formation of the galaxy itself. Galaxies form their stars and their black holes in parallel. How does this construction proceed? Some insight can be gained by directly measuring the evolution of the number of quasars as a function of time, and thereby of redshift. It has long been known that quasars are rare in the nearby universe: we have to look out to distances greater than three billion light years in order to see them. Since intrinsically they are the brightest objects in the universe, it was very soon realized that they must have been numerous in the past (Figure 3.7).

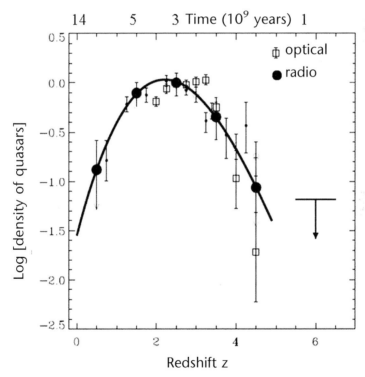

Figure 3.7 Evolution in the number of quasars throughout the history of the universe. The curve represents the density by volume of quasars detected at radio wavelengths (black dots) as a function of redshift z. This curve shows that the period of greatest quasar formation and black-hole growth has been identified (at redshift $z \sim 2$, i.e. between 3 and 4 billion years after the Big Bang). The same is observed of optical quasars (squares), proving that extinction due to dust has little effect (after Shaver et al., 1999).

For a long time, quasars were detectable only in visible light, and it was thought that the rapid falling off in the number of quasars beyond redshift $z = 3$ was due merely to extinction, which becomes ever more marked along the line of sight. Today, thanks to observations at wavelengths (infrared, radio) at which we can pierce the veil of dust, we know that it is not dust that is the problem: the fall-off is real. The number of quasars emitting strongly at radio wavelengths is in the minority (about 10 percent of all quasars), but there is no reason why the proportion should vary with time.

The number of quasars shows a maximum at around $z = 2$, somewhat in the same manner as does star formation. Since the AGN phenomenon tells us at what period mass accumulated in black holes, and caused them to grow, it is now established that the formation of black holes accompanies the formation of stars during the lives of galaxies.

How does a black hole grow?

For today's supermassive black holes to have formed, their mass must have been gathered during a large fraction of the age of the universe, and the original nucleus must already have been large: it is no easy thing for such monstrous objects to accumulate. The earliest stars doubtless gave rise, during the first billion years, to stellar-mass black holes. But how do we make the leap from stellar black holes to galactic black holes? And in such record time, too, given that the earliest quasar is observed at $z = 6.4$, i.e. just after the formation of the first stars, also during the first billion years of the age of the universe?

The major obstacle to the formation of supermassive black holes is that represented by the so-called 'Eddington luminosity limit.' The accretion of matter by the black hole produces a lot of radiation, which reacts with the surrounding matter, repulsing it with radiation pressure. The phenomenon of accretion is therefore a self-regulating one, and the maximum possible luminosity of the black hole is a function of its mass. There is at the same time a maximum amount of matter, the 'Eddington accretion limit,' with which a black hole can be supplied. If the black hole accumulated matter more rapidly, it would become even brighter, and that intense luminosity would exert sufficient pressure to balance out gravity, and prevent extra matter from falling in.

The Eddington luminosity is proportional to the mass of the black hole. Typically, for a supermassive black hole of the order of a billion solar masses, it is of the order of 10^{40} watts, exactly the maximum value observed in quasars. It is therefore thought that the most luminous quasars correspond to black holes of a few billion solar masses, and radiate at the maximum possible rate, i.e. the Eddington luminosity. The maximum rate of matter accretion corresponding to this luminosity, with a radiation efficiency coefficient of 10 percent, is known as the Eddington accretion rate. If it is kept supplied, the black hole can grow indefinitely, although there is a mass beyond which matter will be swallowed without there being time for it to emit much energy, and the galaxy will not be seen as a quasar.

In fact, the radius of the horizon grows with the mass, and the mean density within the horizon decreases as the square of the mass. When this density equals the mean density of the stars, tidal forces are no longer able to destroy them, and they will be swallowed whole by the black hole. Their destruction will occur beyond the point of no return of the light, and the quasar phenomenon will not be seen. This limiting 'Hills mass' has a value of 300 million solar masses. It is easy to calculate the approximate time needed for a black hole to achieve this mass; such black holes are currently observed in galaxies.

One might optimistically suppose, at first sight, that the density of matter at the center of a galaxy would remain more or less constant, feeding the black hole and falling at a velocity similar to the rotational velocity of the matter around the center. If we start with a stellar-type black hole, of the order of 10 solar masses, it would take longer than the age of the universe to attain the Hills mass. It can be shown that the matter accretion rate grows according to the square of the mass, meaning that it accelerates as the object grows. Typically, the accretion rate is inversely proportional to the mass of the black hole: the more massive the black hole becomes, the faster it grows. To attain sufficient masses only a billion years after the Big Bang, the original 'seeds' would already have to be more massive than typical stellar-type black holes.

Note that, during the initial phases, the limiting factor for the growth of the black hole is simply the density of the surrounding matter. But the luminosity, which is proportional to the accretion rate, also increases as the square of the mass. Thus, there will come a time when the 'Eddington luminosity,' which is simply proportional to the mass of the black hole, will be reached. The growth rate will reach its upper limit. This phase, when the active nucleus emits the Eddington luminosity, corresponds to what is thought to be the most visible phase of the AGN. Its lifetime is relatively short, of the order of 40 million years. This is, therefore, the duration of an active quasar cycle.

The first black holes in the early universe and intermediate-mass black holes

For black holes to have a considerable initial mass, it is apparently necessary for them to have been formed from the remains of the first stars, which themselves are supposed to be supermassive. These are known as Population III stars (as explained in Chapter 1, Population I stars are the youngest, and Population II stars are the oldest). Since the lifetime of stars depends upon their mass (and the higher the mass, the more rapidly they evolve), the stars of Population III all disappeared long ago. These stars would have been more than 100 times the mass of the Sun, meaning that they would have had far more mass than the most massive stars in today's universe. They would have formed within small galaxies of less than one million solar masses.

According to this hypothesis, these supermassive stars, not yet containing any heavy elements or dust, had no stellar winds and experienced no mass loss. Beyond 200-300 solar masses, they may even have collapsed directly into black holes without ejecting their outer layers, as present day supernovae do. When the small galaxies thus formed merged, a black hole of 100,000 solar masses may have been left; this was the result of the merger of all the black holes created from the original supermassive stars. Already, at the very beginning of the universe, at one percent of its current age, almost half of all the mass contained within today's massive black holes was present. The rest needed only to grow around these 'seeds.'

The problem we encounter when trying to understand the growth of these intermediate-mass black holes (and we do not yet know if we can observe any today) is that they are not sufficiently massive to be located at the centers of galaxies, and are scattered round and about within them. It is in fact dynamical friction that causes black holes to fall towards the centers of galaxies, and this force is proportional to the mass of the falling black holes. Only if their masses exceed a few million solar masses can we hope to find supermassive black holes at the center of galaxies. So intermediate-mass black holes will follow their orbits within galaxies without encountering each other, and will not merge to form supermassive black holes; this will occur by accretion of the surrounding matter (gas and stars), a process that may be much slower depending upon where these black holes are located.

Even when a binary black hole has been able to form, the merger of the two holes is not automatic, and may take some time. At first the two black holes approach each other due to dynamical friction: their orbital energy causes surrounding stars to be ejected. However, there are soon no stars left near the binary, and only when gas has been accreted, bringing the two black holes near enough to each other, can the emission of gravitational waves take over. If a third black hole encounters the binary, one of the black holes may be ejected into interstellar or intergalactic space.

According to the hierarchical formation scenario, giant galaxies form principally from the mergers of smaller galaxies, on timescales of the order of a billion years. This is accompanied by mergers of supermassive black holes at their centers, if any are present, i.e. in the final phases of their formation. Meanwhile, intermediate-mass black holes are dispersed into the disks of galaxies, and it is very probable that our Galaxy, for example, contains hundreds, or even thousands of them, depending on their mass.

Most of the material in galactic bulges accumulates during galaxy mergers, in the same way as black holes merge, and it is perhaps not surprising that we observe today a correlation between the masses of black holes and those of bulges. As we have seen, the masses of the disks do not figure in this relationship. This may possibly be due to the fact that intermediate-mass black holes are, as we have seen, dispersed within the disk, and have not found their way to the center, there to become both visible and measurable. It has not yet been possible to extend the black holes/bulges correlation to bulges of lesser mass. There ought to be corresponding black holes smaller than a few million solar masses, and therefore intermediate-mass black holes, not confined to the centers of galaxies.

One of the important roles that these intermediate-mass black holes, formed shortly after the Big Bang, might have is to make an effective contribution to the re-ionization of the universe. We know that the formation of the first stars cannot account for all the ionizing ultraviolet radiation, capable of re-ionizing all the diffuse hydrogen gas between the galaxies. The diffuse hydrogen gas that fills the universe recombined at the beginning of the 'dark age,' 380,000 years after the Big Bang, when the cosmic background temperature fell below 3,000 degrees.

Atomic hydrogen absorbs the ultraviolet radiation emitted by stars very efficiently. The radiation from the first galaxies to form at the beginning of the universe could not pierce the intervening fog, hence the term dark age. Nevertheless, the increasing number of stars managed to ionize ever larger 'bubbles' around galaxies, aided by the radiation from the first quasars. When the bubbles came into contact, certain regions of the universe became entirely ionized. It took a billion years for the re-ionization to be completed, as shown in Figure 3.8.

The era of re-ionization lasted for a long time, as observations of the most distant quasars show. In the spectra of certain quasars, observed at distances equivalent in time to more than 90 percent of the age of the universe, there are hydrogen absorption lines so wide that all the radiation from the quasar has been absorbed within a wide frequency domain, corresponding to a wide redshift

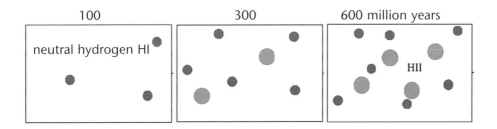

Figure 3.8 Schematic representation of the re-ionization era of the universe. On the right, a region of space almost entirely filled with neutral atomic hydrogen (HI) before re-ionization. The first objects are forming at about 0.6 percent of the age of the universe (100 million years). These objects are dwarf galaxies and their potential wells are very shallow. The equilibrium temperature of the gas in these wells is less than 10,000 degrees, and typically the gas, photoionized by stars, will recombine. But the dense molecular hydrogen that had formed within the wells will be photo-dissociated. In the middle, more massive structures form, when the universe is about 300 million years old. The temperature of these structures is greater than 10,000 degrees, and the ionized gas will be able to start spreading into intergalactic space. On the right, about 600 million years after the Big Bang, the ionized regions (HII) around the galaxies have gained in volume and joined together, thus ionizing large parts of the universe. The end of the re-ionization era is nigh.

82 Mysteries of Galaxy Formation

domain, delimiting the dark age. This discovery, made within the past decade, brought the first direct proof of the existence of the dark age as far away as redshift z = 6.

Binary black holes and the possibility of observing them

Since, according to the current theory of hierarchical formation, galaxies gain mass though mergers, and each galaxy has a massive black hole at its center, each merger must be accompanied by mergers of black holes. It is for this reason that the scientific community is anxious to detect signs of the existence of binary black holes, a phase that ought to precede the merging of black holes. The lifetime of these black-hole binaries is subject to much uncertainty.

It may be that certain manifestations of black-hole binaries have already been observed: for example, the existence of two systems of radio jets, emanating from two interacting quasars, has been observed in the object known as 3C 75. Figure 3.9 shows that the double radio source 3C 75 is evolving within a cluster of galaxies (Abell 400). Two galaxies in this cluster are in the process of merging, and each contains a supermassive black hole. The gas falling into these black holes causes the ejection of jets of plasma, radiating in the radio domain as a result of the 'synchrotron mechanism' (radiation of accelerated charged particles in a magnetic field). The cluster Abell 400 is bathed in a huge quantity of hot gas (at about ten million degrees) emitting X-rays. The plasma jets have been subject to dynamical pressure from very hot gas, by the fact of the rapid motion of the galaxies within the cluster. It might be said that the jets have the wind in their faces! The similarity in the shapes of the jets from the two galaxies proves that they are 'running together' through the cluster, and that they will soon merge. Computer simulations indicate that this galactic merger (and the merger of the corresponding black holes) could occur in less than 100 million years.

Also, there are certain radio jets that are evolving in a helicoidal shape, showing precessional configurations characteristic of the rotation of one of the black holes of a binary system. A very regular variability corresponding to the rotation period of a binary black hole has also been studied for nearly a century in a very unusual object known as OJ 287. This series of observations (Figure 3.10) shows double peaks of intensity with a characteristic period of 11–12 years. The motion of OJ 287 corresponds to Keplerian orbits, and the mass that has been derived for each black hole is some hundreds of millions of solar masses. About ten similar objects, with periods of optical variability of between two and twenty years, are good candidates for binary black holes, and their expected lifetime will be no longer than a few hundred million years.

The origins of black holes 83

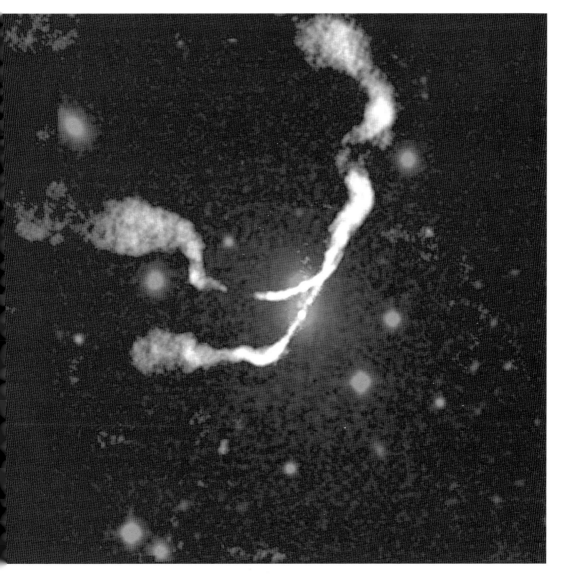

Figure 3.9 Double radio source, two pairs of jets and a binary black hole in the process of formation. The radio source 3C 75 consists of the synchrotron radio emission of the radio jets issuing from two galaxies at the center of the cluster Abell 400. The two pairs of jets originate in the black holes in the nuclei of the two central galaxies. These galaxies are moving at high speed through the cluster, which is filled with very hot gas emitting X-rays. This is the equivalent of an intergalactic wind which is curving the jets back rather in the manner of a scarf worn by a running person (NRAO). See also PLATE 11 in the color section.

84 **Mysteries of Galaxy Formation**

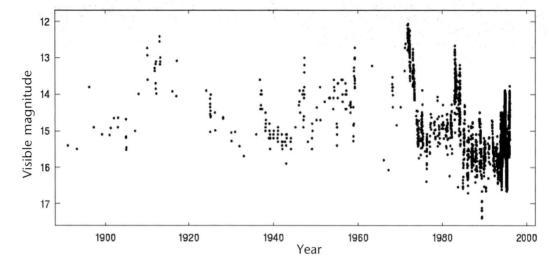

Figure 3.10 Light curve of the quasar OJ 287, in visible light. The quasar is doubtless identified with a binary black hole, and the rotation period of one black hole around the other is detected in the variations in light from the quasar. This historic curve records brightness variations for a whole century. In the recent evolution, we see periodic spikes; the period is 11.86 years (after Pursimo et al., 2000).

The observation of binary black holes: clues to the demography of black holes?

The fact that the masses of black holes are proportional to that of galactic bulges seems to show that the growth of the black holes is synchronized with that of the bulges, or at least that they gather mass as a result of the same events, for example in the course of mergers. If a merger of two galaxies produces a binary black hole, the too-rapid arrival of a third galaxy might produce a third body at the center of the grouping. However, a three-body system is unstable, and one of the black holes runs the risk of being ejected. Such ejections can certainly not be frequent, since they would compromise the simultaneous growth of the bulge and the black hole.

Nevertheless, the risk of ejection resulting from the accretion of a third body is high. The third body is very likely to come on the scene: the accretion of a third galaxy is a frequent occurrence, since galaxies exist in groups. The lifetimes of binary black holes must be short enough to forestall ejections as a result of the 'gravitational slingshot' effect. Even today, the mechanisms which could reduce the lifetimes of binary black holes are very poorly understood. At first, the two black holes spiral towards each other, losing orbital energy as they eject stars from the center of galaxies. Now, when the black holes have created a void around themselves, there are no nearby bodies left to drain their energy and angular momentum. It is thought that the binary black hole is not in fact stabilized at the center of two galaxies, but experiences random oscillations during the course of which it encounters more stars and coalesces more rapidly. Moreover, pulses of interstellar gas regularly fall towards the center, and their accretion is another factor in the drawing together of the two black holes. Finally, the lifetime of pairs of black holes is likely to be less than the mean duration of two galaxies merging, which would prevent the ejection of supermassive black holes into intergalactic space, to be lost forever to the system.

Note that binary black holes are intensively sought in order to test theories of gravity in strong fields, and also as potential emitters of gravitational waves, which may soon be intercepted by new generations of detectors, although none has yet been directly detected.

Activity in black holes: 'downsizing'

A paradoxical phenomenon involved in the scenario of the hierarchical formation of galaxies is also found with black holes. While the dark halos of galaxies can only grow over the course of time, the formation of stars and phenomena of reaction and suppression act to stop the formation of stars in the largest galaxies which must, therefore, have formed very early on – at least eight billion years ago. Thus it is that today, only small galaxies can form.

The observation of the active nuclei of galaxies reveals a similar scenario.

Very luminous AGN phenomena, associated with the most massive black holes, all took place very early on, only three billion years after the Big Bang, as shown in Figure 3.11. Also, the activity in nuclei that we observe locally essentially involves small black holes: for example, those associated with Seyfert galaxies (spiral galaxies with small bulges), with black holes of between 10^6 and 10^8 solar masses.

This is a paradoxical phenomenon, which seems to contradict the scenario of the hierarchical formation of structures, within which the smallest form first and eventually merge to form large ones. How do we explain this paradox?

The hierarchical formation of structures in fact involves dark matter, i.e. the halos that surround galaxies. These halos progressively merge, though no energy is dissipated, since dark matter undergoes no collisions. There is no other phenomenon to hinder successive mergers, and galactic-sized halos merge to form groups and even clusters of galaxies. Within these great structures, the individual halos of galaxies gradually lose their identities, merging into one supermassive halo involving the whole cluster.

This is not the case with visible (baryonic) matter in galaxies: visible galaxies retain their identities within clusters. Relative velocities within clusters are so great that interactions between galaxies are no longer effective, and mergers no longer take place. What is more, the interstellar gas in galaxies is swept away into the cluster: as it enters the cluster at high speed, the dynamical pressure of the gas already at rest within the cluster acts like a high-velocity wind, ejecting gas

86 Mysteries of Galaxy Formation

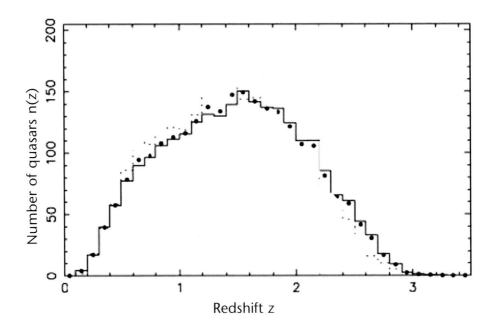

Figure 3.11 Distribution in redshift of the 23,000 quasars in the 2dF catalogue, whose spectra were obtained using the 4-m Anglo-Australian Telescope, by Croom et al. (2004). Note the peak in the number of quasars between redshifts $z = 1$ and $z = 2$.

from within the galaxies. These galaxies no longer have a supply of fresh gas, since all the intergalactic gas had been heated to high temperatures of several tens of millions of degrees when the cluster was formed. Collisions between galaxies will therefore no longer lead to a considerable dissipation of energy.

The first dark halos formed at the beginning of the universe are denser than those found today. This is due to expansion, which causes the mean density of the universe to decrease. For a structure to decouple from the expansion, its density must be at least twice as great as the mean density of the universe: so the structures formed are less and less dense as time passes. A high density shortens the time needed for dynamical evolution. Everything happened more quickly in the beginning.

Moreover, the quantity of gas available in galaxies was much greater at the beginning of the universe, when stars had not yet all formed. Given that clusters of galaxies had not yet formed, the accretion of gas from cosmic filaments outside galaxies was still proceeding apace. All the circumstances for building black holes, and for their rapid growth, were then present. Even though the dark halos were not very massive, the visible galaxies could already have become massive, in the same manner as the supermassive black holes in their centers. However, the great efficiency of the processes evidenced by the growth of the mass and ready supply of material to black holes would very soon be diminished, and the density of very luminous AGN would fall off rapidly.

Of course, the massive galaxies formed during this very active era in the early universe are still with us today, but they are found in superdense regions of the universe which have seen the development of clusters and superclusters of galaxies. Their gas supply is soon almost exhausted, and they then evolve passively, as their stars age, and they accumulate only a negligible quantity of gas.

The formation of galaxies and black holes now persists only within the voids between clusters. However, expansion has reduced densities and prolonged the time required for dynamical activity; any formative activity proceeds more slowly, and less gas is available to form galaxies. The galaxies which form today are of small mass and exhibit low activity, in both star formation and the growth of nuclei.

In addition to this general scheme, there is the fact that mergers between galaxies markedly increase the rate of star and nucleus formation, by causing gas to fall into the center of galaxies. The most favored era for interaction between galaxies occurred during the first part of the lifetime of the universe, when groups and clusters of galaxies were forming, about ten billion years ago. The size of the structures decoupling from the expansion can only grow with time, and today this is the case with superclusters. The majority of mergers took place during the epoch when groups and clusters were formed, thanks to encounters between galaxies within vast structures of dark matter. The cluster having formed and 'relaxed,' the gas supply to the galaxies ceased and the black holes at their center were starved. Today the number of mergers has fallen by a factor of about 20, compared with its maximum in the past. Figure 3.12 shows how galaxy formation models explain the more efficient formation of massive galaxies and very luminous AGN preferentially at the beginning of the universe.

Self-regulation phenomena

The significant correlation between the masses of black holes and those of galactic bulges brings with it a certain number of questions that have not yet been resolved: should there not be situations where one grows more quickly than the other, or where the black hole is ejected and the correlation fails? For example, there exist rapid increases in stellar formation in cases where the stellar mass of a galaxy increases considerably within the nucleus, while the black hole is not supplied. This means that there will be delays in the growth of the black hole, which will have to be compensated for at a later phase.

Remember that the mass of galactic disks is not involved in this correlation: they definitely lack a central potential well that would allow intermediate-mass black holes to accumulate, and these stay far from the center, dynamical friction being insufficient to draw them into the same region. As galaxies interact, these black holes can be 'peeled away,' together with part of the galactic disk. Only bulges will be able to draw in and accumulate black holes.

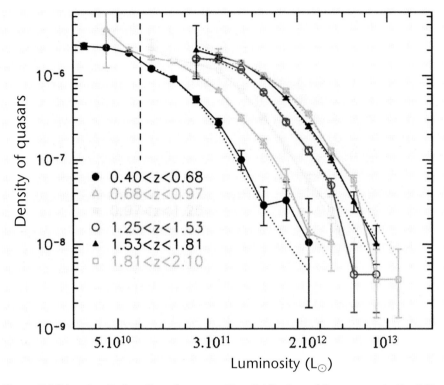

Figure 3.12 Luminosity function of quasars. The distribution of the quasars in the 2dF catalogue as a function of their luminosity for the six intervals of redshift indicated on the diagram. For each redshift interval, the number of quasars falls off spectacularly, and almost exponentially, in the direction of the highest luminosities. The dotted lines represent the best model fitting the observations. The various curves are very much the same, with higher redshifts corresponding to the upper curves, i.e. to higher densities, especially at high luminosities. At high redshifts, luminous quasars are therefore more numerous. In the models, the growth of supermassive black holes and their quasar radiation occurs essentially during galaxy mergers. The decrease in the number of quasars with time is therefore due to the reduction in the number of galaxy mergers; but it is also due to the decrease in the density of the gas, and to the lengthening of the dynamical time (see text for more details).

Observations have been made which suggest that the concomitant growth of black holes and galaxies is liable to certain exceptions. For example, among galaxies with active nuclei, Seyfert galaxies are normally characterized by very wide emission lines from their nuclei. However, a certain class of galaxy has recently come to light: the 'narrow-line Seyfert galaxy.' There are several indicators suggesting that these galaxies have lighter central black holes compared with other galaxies with active nuclei. On the one hand, there is the narrowness of the spectral lines, so typical of these objects: since we already know the size of the region within which these lines are produced (and measured by methods involving the reverberation of light), the mass of the central black hole can be deduced by methods involving dynamical equilibrium. On the other hand, the distribution of energy as a function of wavelength observed in these particular active nuclei is not what might be expected from an accretion disk around a very massive black hole.

The accretion rate of black holes in these narrow-line Seyfert galaxies seems to be much greater than is normal. In fact, they are very luminous, although of relatively low mass. The speed at which the black hole grows, which can be roughly calculated by dividing the mass of the black hole by the accretion rate, is very great. In other words, these objects appear to be young active nuclei with black holes in rapid growth. In these conditions, it is easy to see that these galaxies are apparently not obeying the same relationship between the mass of the bulge and that of the black hole as observed in other galaxies; their black holes seem to be 'underweight' when compared with their bulges, which have already grown. Is this phase, identified by means of the narrow lines, the main phase in the history of the cosmic growth of black holes?

Statistically, the comparison between the low number of galaxies with an active nucleus, and all the others with a 'quiet' central black hole, indicates that the 'active' phase of a galaxy's life is very brief. The duration of this activity, and thereby of that of the accretion of matter, is estimated to be around 100 million years. Among these active nuclei, a fraction of between 10 percent and 30 percent would now be in the phase of very rapid accretion. In this phase the mass of the black hole grows by a factor of 1,000! For the remainder of its active period, the black hole grows much less vigorously: this might be termed the 'old age' of the black hole.

The census of active nuclei and the statistical data relating to their accretion rates have been established in any detail only for relatively nearby objects, corresponding to the second half of the age of the universe, going back seven billion years. During the first half it seems very likely that the growth of black holes was much more rapid than it is today.

And what if the opposite were true?

Finally, it is possible that, on the contrary, it is black-hole activity that regulates the formation of stars within galaxies. Once activity has been triggered in the nucleus, very energetic phenomena emit radiations that heat and ionize the surrounding gas, arresting its cooling and condensation into stars. This self-regulation may remain confined to the center of the galaxy, but may also propagate further out, thanks to plasma jets emitted by the nucleus or the accretion disk. These jets, of which we have seen an example in Figure 3.9, may extend well beyond the radius of a galaxy. Also, they are not always channeled in the same direction, but, as a result of precession due to various phenomena, such as that of binary black holes, or that of dynamical instability associated with

90 Mysteries of Galaxy Formation

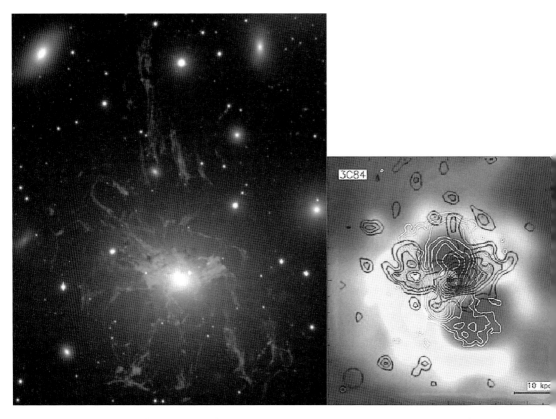

Figure 3.13 Self-regulation phenomena in NGC 1275 (Perseus A), the central galaxy of the Perseus cluster. **Left:** a deep hydrogen-alpha image of NGC 1275, taken by the WIYN 3.5-m telescope at Kitt Peak National Observatory. The filaments emanating from this galaxy are probably the result of an interaction between the black hole in the center of the galaxy and the intracluster medium surrounding it. The glowing background objects in this image are galaxies in that same galaxy cluster. At a distance of about 230 million light-years, this is the nearest example to Earth of such vast structures, which are seen surrounding the most massive galaxies throughout the Universe (C. Conselice/Caltech and WIYN/NOAO/AURA/NSF). **Right:** an image of the X-ray emission from the very hot gas in the cluster (Chandra X-ray Observatory, Fabian et al., 2000). The contours shown in white represent the emission from radio jets (Pedlar et al., 1999). The cavities created by the radio jets in the hot gas of the cluster are very well seen in the right-hand image. The plasma ejected by the active nucleus at the center of the galaxy produces bubbles of diffuse gas which rise through Archimedean pressure, finally falling back onto the galaxy, as filaments of cooler gas. See also PLATE 12 in the color section.

accretion disks, the jets travel intermittently in random directions, and sometimes even along the plane of the galaxies!

Figure 3.13 illustrates these self-regulation phenomena due to radio jets from the radio source Perseus A, which considerably perturbs the gas around the central galaxy, NGC 1275. These phenomena can perhaps partly explain the correlation between the mass of black holes and the mass of stellar bulges within galaxies.

In conclusion...

Throughout the lifetimes of galaxies, mass accumulates at their centers, leading progressively to a more or less massive bulge, or a supermassive black hole. These black holes merge not long after the merger of the parent galaxies. The infall of matter onto the black hole, which causes it to grow, is very rapid, and represents but a very short phase, of the order of just a few percent of the life of a galaxy. This figure also represents the probability of finding an active nucleus in a galaxy. Self-regulation phenomena closely link bulges and black holes: black-hole activity can also arrest the importation of mass and the formation of stars at the center.

4 Scenarios of galaxy formation

How are galaxies formed? The essential factors are: the rate at which the masses of gas collapse inwards; the rate of rotation, which tends to form disks; the efficiency of star formation; and above all, the environment, the richness of which depends upon initial large-scale density fluctuations.

There are many scenarios which can account for the formation of galaxies: monolithic collapse; the violent growth of galaxies by mergers (the 'hierarchical' scenario); or more moderated growth through secular evolution. All these have their more or less important parts to play in the course of cosmic evolution.

Rhythms of galactic growth are very varied, depending on the environment: the largest galaxies grow fastest, within clusters, and soon lose their gas. The life expectancies of galaxies are also very varied. Today, the most massive of them are already dead, in the sense that they no longer harbor the formation of new stars.

The formation of structures: 'top-down' or 'bottom-up'?

Currently, the most commonly agreed scenario of the formation of structures is the hierarchical, according to which the first structures to form were the smallest. These then merged to form larger ones.

This has not always been the case: in the 1970s and 1980s, two completely opposing models were current:

- the 'top-down' model held that the largest structures, clusters and superclusters formed first, and then broke up into galaxies;
- the 'bottom-up' model held that small structures formed first and then merged to form larger ones. It is this model which has been developed into today's hierarchical scenario.

What differentiated these two models was their view of the nature of the primordial fluctuations from which structures developed. Various modes of fluctuation, in combination or in isolation, may be envisaged for the beginning of the universe:

- the existence of density fluctuations corresponding to fluctuations in gravitational potential, which include all particles, including photons. They

are accompanied by fluctuations in temperature and pressure such that constant entropy is maintained. It is for this reason that we call this type of fluctuation 'adiabatic.' In these fluctuations the number of photons remains proportional to the number of particles of matter;
- in the isothermal mode, photons do not follow matter, and do not fluctuate. The temperature therefore remains constant. This mode was proposed mostly before the effective measurement of anisotropies in the cosmic background radiation.

Since we have observed temperature fluctuations in the cosmic microwave background, the most widely considered 'non-adiabatic' models are the 'isocurvature fluctuation' types which conserve curvature and within which the mass is uniform. The reality could be a combination of these two types of modes, independent of fluctuations. However, observations suggest that adiabatic-fluctuation models are more likely.

We must also remember that each type of fluctuation is associated with a type of dark matter. This may exist in the form of cold dark matter (CDM) or hot dark matter (HDM). Their 'cold' or 'hot' character is based on the criterion of the mean velocity of the particles when they decoupled from the primitive plasma, where photons, baryons and dark matter were in equilibrium.

Photons decoupled from the primitive plasma 380,000 years after the Big Bang. Neutrinos had decoupled even before the photons and have long remained relativistic, i.e. moving at a velocity close to that of light. Their velocity will be significantly reduced only later, depending on their mass. Neutrinos are therefore the best candidates to be constituents of hot dark matter. The consequence of their relativistic velocity is that these 'too-hot' particles will prevent small-scale fluctuations in matter from concentrating together under their own gravity.

In fact, neutrinos, like any other HDM candidates, do not interact with matter, and travel for distances similar to those travelled by photons: they have a mean free path almost to the horizon, and their 'pressure' stabilizes matter on these scales. Only on greater scales will there be collapse under the influence of gravity. This phenomenon, inherent in all HDM candidates, explains the great difference between models of the formation of structures: in HDM models, large structures collapse first, and fragment to form galaxies. In the CDM models, the reverse is true.

'Top-down' models, within which large structures appear initially and then collapse, are known as 'pancake' models. They offer a natural explanation of the planar/filamentary structure of the universe, since the collapse occurs on a grand scale when the pressure of matter and baryons is negligible.

Let us now imagine a field of random fluctuations and random density distribution. If the probability of collapse in a particular direction is, for example, 0.5, then the probability of collapse in three axes simultaneously is only 0.125. In fact the collapse is non-linear and accelerates very quickly. Matter will become very dense in this direction, precipitating the collapse, which then develops

Scenarios of galaxy formation 95

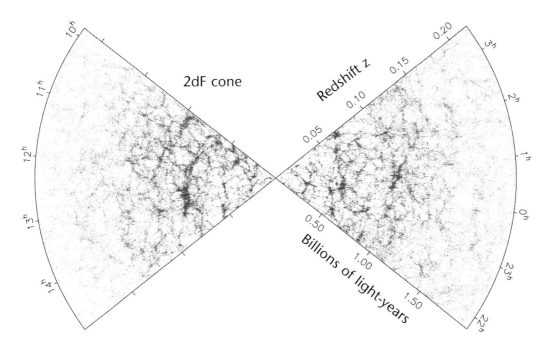

Figure 4.1 Three-dimensional survey of the large-scale structures of the universe. In this 2dF (2-degree Field) survey, in addition to the two dimensions of the plane of the sky, there is a third dimension obtained by measuring the velocities or redshifts of 221,000 galaxies. The 2dF survey, by the Anglo-Australian Telescope (AAT), is based on areas of the sky measuring two square degrees. Each dot on this chart is a galaxy, whose distance in billions of light years is inferred from its radial velocity. The survey is complete, out to a distance of 1.5 billion light years. Galaxies are distributed in clusters and superclusters, along a filamentary structure around bubble-like voids, the whole reminiscent of a cobweb. See also PLATE 13 in the color section.

preferentially into a flat 'pancake' structure. These planes will later become superclusters of galaxies. Where the planes intersect, we may find dense filaments, and at the junctions of the filaments, clusters of galaxies.

When we contemplate the large structures of the universe traced out by the galaxies (see the 2dF survey and Sloan Digital Sky Survey (SDSS), Figures 4.1 and 4.2), or even the results of numerical simulations (Figure 4.3), we cannot help but be struck by the filamentary structure, resembling a network of interconnected threads, within which is plenty of empty space. The most extensive structures, called 'Great Walls,' are of a size comparable to that of the mapped regions themselves, or some few percent of the extent of the horizon (to ∼500 Mpc, Figure 4.2).

96 Mysteries of Galaxy Formation

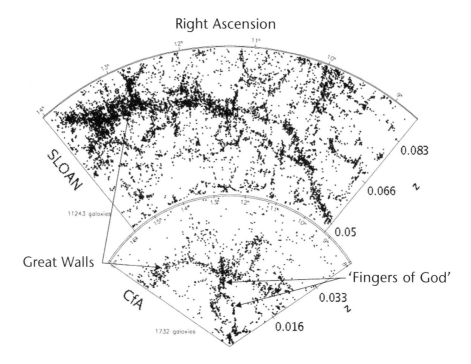

Figure 4.2 Great walls of galaxies in cross-section. In this figure, two slices of the universe are compared, the upper from the Sloan Digital Sky Survey (SDSS), covering 4 degrees, and the lower from the Center for Astrophysics (CfA), covering 12 degrees. The 'great walls of galaxies' become ever bigger as the various surveys progress, and are as wide as the slices of the universe explored (after Gott et al., 2005). In the CfA slice, we see radial structures, a phenomenon involving artificial filaments extending towards the observer. These structures are artifacts associated with the particular velocities of galaxies superimposed on expansion, and are known as 'Fingers of God.' These particular 'fingers' are in the Coma Cluster of galaxies.

The formation of structures by mergers

Observation of the parameters of the universe, and especially of the cosmic microwave background, have confirmed that we are indeed in a 'bottom-up' cosmos where the first structures to be formed were small.

The paradigm for dark matter adopted by most astrophysicists is that of cold dark matter (CDM), which supplies the basis for the hierarchical scenario of the formation of dark-matter structures (Figure 4.4).

Since the physics of dark matter is very simple, as it does not interact with itself or any other matter other than gravitationally, the formation processes are relatively well understood. Numerical simulations involving only dark matter are

Scenarios of galaxy formation 97

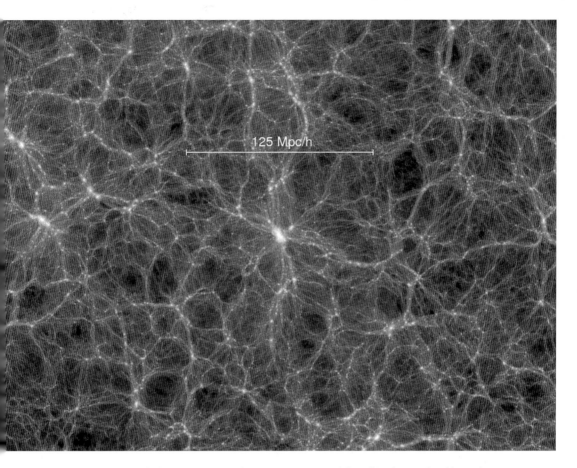

Figure 4.3 Detailed computer simulations are a powerful tool in the study of how the galaxies we see in the universe today formed out of the small fluctuations in matter density that were present in the primordial universe. To model this highly non-linear and intrinsically three-dimensional process, the matter fluid may be represented by a collisionless N-body system that evolves under self-gravity. However, it is crucial to make the number of particles used in the simulation as large as possible to model the universe faithfully. Scientists at the Max-Planck-Institute for Astrophysics, together with collaborators in the international Virgo consortium, carried out a new simulation of this kind using an unprecedentedly large number of more than 10 billion particles. This is an order of magnitude greater than the largest computations previously carried out, a fact that inspired the name 'Millennium Simulation' for the project. In this single frame, a slice through a small part of the simulated universe, the large-scale structure in the dark matter is seen. Clearly visible is the 'Cosmic Web' which connects individual galaxies, groups and clusters by filaments of dark matter, surrounding large underdense voids (Max-Planck-Institut für Astrophysik and the Virgo consortium). See also PLATE 14 in the color section.

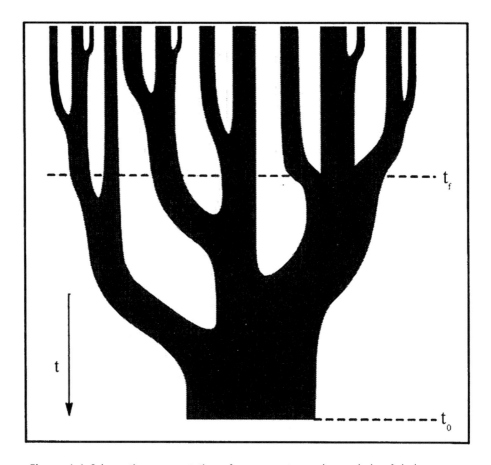

Figure 4.4 Schematic representation of a merger tree, where a halo of dark matter corresponding to a galaxy such as the Milky Way (the lower trunk of the tree) forms from successive mergers of smaller halos. Time runs from the top to the bottom of the diagram, and the width of the branches represents the mass of the various halos merging to create the final halo. The horizontal dashed lines show the present time (t_0) and the time of formation (t_f), defined as the time at the end of which a parent halo is formed containing at least half the mass of the final halo (after Lacey and Cole, 1993).

sure successes, especially at large scales. Simulations are indispensable as soon as the perturbations become non-linear, which occurred very early on in the history of the universe in the case of small structures (galaxies formed within the first billion years), but also later in the case of clusters and superclusters, which are still forming today.

The difficulty arises when we consider baryons, i.e. ordinary matter, stars and interstellar gas, within which physical processes are extremely complex: an unfortunate state of affairs, since this is the only matter which is visible and can

Scenarios of galaxy formation

be used to match the theory with the observations. The standard model holds that, just after the recombination of matter (380,000 years after the Big Bang), neutral gas was able to collapse into gravitational potential 'wells' already created by dark matter. In this context, the accumulation of matter in galaxies could occur in at least two ways:

- matter streaming along filaments 'feeds,' by accretion, the dark-matter halos acting as 'wells' along the length of the filaments and at their junctions;
- galaxies interact with one another, losing their relative orbital energy and merging. Successive coalescences are one way of constructing giant galaxies.

Mergers are very efficient all the while the relative velocities of the galaxies are of the same order as their internal rotational velocities, i.e. while they do not belong to a much more massive, larger structure, for example a cluster of galaxies.

Today, galaxies are still interacting and merging (Figures 4.5 and 4.6), especially in the least populated regions, outside rich clusters of galaxies, but at a rhythm that has been much reduced over time. Certainly, the golden age of interactions was when the clusters were forming – when relative velocities were not yet too high, and in dynamic equilibrium with the gravitational depth of the clusters.

Several scenarios for galaxies

The formation of the visible galaxies presents two problems:

- when did the component stars form?
- when was the main mass assembled?

In fact, these two questions can be dissociated, for it is possible that there are galaxies merging today which have no interstellar gas, and therefore no stars form in them. The history of the formation of stars can be reconstituted for nearby galaxies, in which individual stars can be resolved whose magnitudes, colors and other characteristics are known. The technique of dating stellar populations by their colors can be broadly applied to systems in which individual stars cannot be resolved. However, this technique has its limitations, not least because of confusion between, for a given color, stars' ages and metallicities; another factor is the lack of separation of important age-groups, both spatially and in time, the colors of old populations being quite similar, after a few Gyrs.

One of the first scenarios of the formation of spiral galaxies such as the Milky Way was proposed in the 1960s. It involves the progressive collapse of a sphere of gas under its own gravity. As it contracts, its rotational velocity increases, obeying the law of the conservation of angular momentum. The collapse is rapid, since the gas is almost in free fall, but the rotation does have some slowing effect. The collapse occurs essentially in the same direction as the axis of rotation, and

Figure 4.5 This image from the Hubble Space Telescope's Advanced Camera for Surveys shows the Antennae galaxies – seemingly a violent clash between a pair of once isolated galaxies, but in reality a fertile marriage. As the two galaxies interact, billions of stars are born, mostly in groups and clusters of stars. The brightest and most compact of these are called super star clusters. The two spiral galaxies started to fuse together a few hundred million years ago making the Antennae galaxies the nearest and youngest example of a pair of colliding galaxies. Nearly half of the faint objects in the Antennae are young clusters containing tens of thousands of stars (NASA, ESA, the Hubble Heritage Team (STScI/AURA), and the ESA/Hubble Collaboration, with thanks to B. Whitmore (STScI) and James Long (ESA/Hubble)). See also PLATE 15 in the color section.

Scenarios of galaxy formation 101

Figure 4.6 A spectacular pair of interacting galaxies captured by the Hubble Space Telescope's Advanced Camera for Surveys. Located 300 million light-years away in the constellation Coma Berenices, the colliding galaxies have been nicknamed 'The Mice' because of the long tails of stars and gas emanating from each galaxy. Computer simulations reveal that we are seeing two nearly identical spiral galaxies about 160 million years after their closest encounter. The long, straight arm is actually curved, but appears straight because we see it edge-on. The simulations also show that the pair will eventually merge, forming a large, nearly spherical galaxy. The stars, gas, and luminous clumps of stars in the tidal tails will either fall back into the merged galaxies or orbit in the halo of the newly formed elliptical galaxy (NASA, H. Ford (JHU), G. Illingworth (UCSC/LO), M.Clampin (STScI), G. Hartig (STScI), the ACS Science Team, and ESA). See also PLATE 16 in the color section.

the gas flattens out into a disk. This process is slow enough for stars to be able to form as the gas spreads into a disk. The first stars therefore assume a spheroidal arrangement, with little rotation, and their morphology corresponds to the degree of flattening of the gas at the epoch of their formation. This scenario predicts that the further away from the disk a star is located, the older it will be. This seems to be true for most stars.

The scenario was proposed to take account of the stellar halo of the Galaxy, in which the abundance of metals in stars seems to be related to the eccentricity of their orbits and to their distance from the galactic plane. It is only logical that the enrichment of the Galaxy in heavy metals, as nucleosynthesis takes place inside stars and stellar winds and explosions distribute its products, will be pursued as new generations of stars form nearer to the disk. We can confirm that, currently, young stars are being formed within the thin disk itself (Figure 4.7).

However, this relatively rapid collapse can no longer account for all the observations of the current Milky Way, and its applicability to the formation of

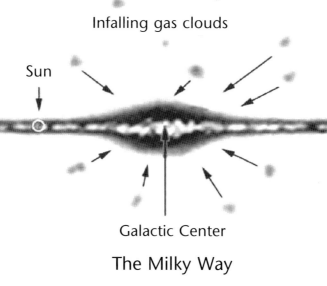

Figure 4.7 Representation of the Milky Way, seen edge-on, with the Sun located in the outer part of the disk of stars. The dark band represents dust, the signature of the young, thin disk. In the monolithic collapse scenario, the initial sphere of gas clouds collapses in free fall, gradually forming the disk, and the stars formed during the collapse globally conserve their spheroidal morphology and constitute the stellar halo, where metallicity is related to the eccentricity of the orbits.

our Galaxy is in doubt. The current size of the disk is much greater than that of the stellar halo. However, it could be that this collapse might have played a role in the formation of the central bulge. More generally, this 'monolithic' rapid-collapse theory was widely taken up to explain the formation of spherical and elliptical galaxies. This mechanism is often evoked to take account of the observation at large redshifts (i.e. during the first billion years of the history of the universe) of massive elliptical galaxies already at the end of their evolution. This scenario concurs with that of the hierarchical formation of galaxies through interaction and merger, which will now be described. The accretion of satellite galaxies was evoked early on (from the 1970s) to explain certain discordant observations in the halo of the Milky Way. Globular clusters, for example, are ancient stellar systems with very low abundances of metals, and where the abundances appear to be unrelated to their distance from the center (Figure 4.8).

The random accretion of a series of protogalaxies, and the later accretion of gas to form the younger disk, was a scenario which furthered our understanding of the observations. Today, this scenario is gaining more ground as we discover more and more streams of stars of low metallicity around the Milky Way, these are dynamically coherent, and are identified as tidal streams caused by the

Scenarios of galaxy formation 103

Figure 4.8 M80 (NGC 6093), one of the densest of the 147 known globular star clusters in the Milky Way galaxy, captured by the Hubble Space Telescope's Wide Field and Planetary Camera 2. Located about 28,000 light-years from Earth, M80 contains hundreds of thousands of stars, all held together by their mutual gravitational attraction. Globular clusters are particularly useful for studying stellar evolution, since all of the stars in the cluster have the same age, but cover a range of stellar masses. Every star visible in this image is either more highly evolved than, or in a few cases more massive than, our own Sun. Especially obvious are the bright red giants, which are stars similar to the Sun in mass that are nearing the ends of their lives (NASA, the Hubble Heritage Team (STScI/AURA), with thanks to M. Shara and D. Zurek (AMNH), and F. Ferraro (ESO)). See also PLATE 17 in the color section.

104 **Mysteries of Galaxy Formation**

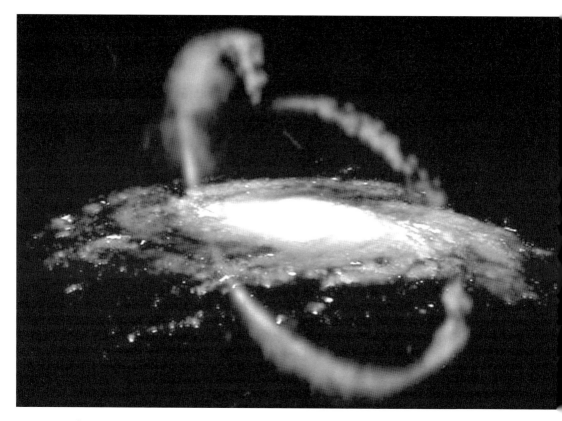

Figure 4.9 Artist's impression of our Galaxy, the Milky Way, surrounded by a stream of stars. Numerous stellar streams and tidal debris have recently been discovered around and in the halo of the Milky Way. This artist's impression shows a star stream recently 'peeled' off the Sagittarius dwarf galaxy. This stream will long pursue the dwarf galaxy's former orbit around the Milky Way. Several similar streams of tidal debris, of low metallicity, have been identified, showing that our Galaxy is still in the process of swallowing numerous small neighboring galaxies. See also PLATE 18 in the color section.

destruction of dwarf companion galaxies of the Milky Way. The dwarf galaxy in Sagittarius is a very representative example of such galaxies (Figure 4.9), as is the Canis Major dwarf galaxy, among others. It could even be that the stellar halo of our Galaxy is entirely made up of the debris of neighboring galaxies absorbed at different epochs. This process of aggregation of mass was certainly more important in the past, given the higher rates of interaction and merging that we see at increasing cosmic distances.

A comparison of the two processes of formation already mentioned (monolithic collapse and the hierarchical scenario) is schematically shown in Figure 4.10. In the former case, star formation takes far less time than the duration of the collapse and the stars form spheroidal systems and elliptical galaxies. Spheroids or bulges form before the disk; whereas, in the hierarchical

Scenarios of galaxy formation 105

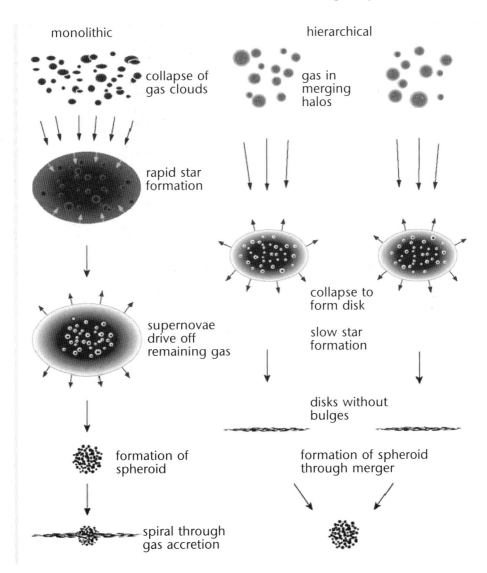

Figure 4.10 The two main scenarios of galaxy formation: monolithic collapse and hierarchical formation through mergers. **Left:** a group of gas clouds collapses, almost in free fall, under its own gravity. The time during which cooling and star formation occur is comparable to the collapse time, of the order of hundreds of millions of years, and stars begin to form into a spheroidal system even before the gas has had time to flatten out into a disk. After contraction, and the expulsion of gas through stellar winds and explosions, an elliptical galaxy or a spheroidal bulge is formed after one or two billion years. This monolithic scenario requires later accretion of gas to form the disks of spiral galaxies. **Right:** the gas collapses more rapidly, and the cooling/star formation period is longer. The gas has time to flatten into a disk, within which stars later form. In this scenario, the disks form before the bulges. The formation of spheroidal systems or elliptical galaxies is now the result of mergers of spiral galaxies (after Ellis et al., 2000).

106 Mysteries of Galaxy Formation

scenario, disks form first, and only then are spheroids formed through later mergers of two or more spiral galaxies.

The secular evolution of galaxies

Alongside these two scenarios, we find another: secular evolution, involving an internal dynamical evolution of galaxies fed by regular accretion of matter from cosmic filaments in their vicinities. This scenario is schematically summarized in Figure 4.11. Gravitational instabilities provide the motor for the internal dynamical evolution, forming the spiral arms and bars within disks.

Bars are density waves which make the disk non-axisymmetrical. There are therefore tangential forces which act upon the gas and lead to change in the angular momentum. When a robust stellar bar has been formed because of instability in the disk, the interstellar gas will lose its angular momentum and fall towards the center. The immediate effect is a flurry of star formation. This is confined at first to a ring around the nucleus where the gas accumulates, as a

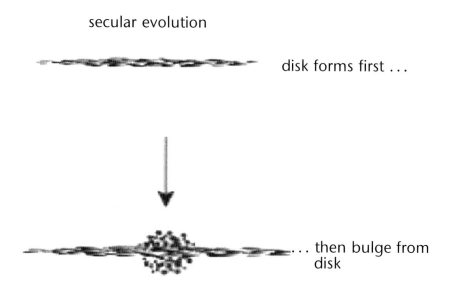

Figure 4.11 The secular evolution scenario. A disk of gas and stars is unstable, and forms spiral density waves and bars. These asymmetries cause the matter to fall towards the center. Through resonance effects between the bar and the stars, the latter are raised above the plane of the galaxy and can form a central bulge. So it is possible that certain bulges have been formed from disks. These processes can be maintained as long as the galaxy is acquiring matter and gas from outside.

result of dynamical resonance with the bar. This characteristic phenomenon gives rise to a number of bright rings of young stars in barred galaxies. Matter gradually accumulates at the center, so the bars are a means of concentrating mass. However, this is a self-regulating mechanism: as the gas falls inwards, it transmits its angular momentum to the bar, which weakens it and slowly destroys it. Only if another disk, richer in gas, is reconstituted through accretion from outside will a new instability, in the form of a barred density wave, be able to continue the process.

The frequency of galactic bars gives an insight into the importance of the phenomenon. In atlases of spiral galaxies photographed in visible light, the fraction of barred spirals is some two-thirds (with about one-third having robust bars and about one-third weaker bars). However, this is to underestimate the frequency of bars, because dust, which absorbs radiation in the visible domain, also accumulates around the center. Currently, near-infrared images, where the effect of the dust is considerably reduced to give a truer view of the mass of the stellar component, show that about 80 percent of today's galaxies are barred.

This high frequency shows that the renewal of bars is a very efficient process in spiral galaxies. The amount of gas that a galaxy must acquire to maintain these bars is considerable: enough to double its mass in a period less than that of the Hubble time[1]. The frequency of bars reveals the importance of the secular evolution theory, which certainly acts in tandem with the other two scenarios (Figure 4.12). What is the frequency of bars in the course of time? Were the first galaxies of the early universe barred? The question is a tantalizing one, since to answer it demands a sensitivity and image quality of our telescopes which they simply do not possess when we observe very distant galaxies.

The first astronomers to address this problem (too) soon concluded that bars were not present in objects with large redshifts. Images in 'visible light' were in fact in the ultraviolet for distant galaxies, and even today bars are invisible at ultraviolet wavelengths. However, working in the near-infrared, which corresponds to visible light as far as distant galaxies are concerned, and comparing the frequency of only those bars large enough to be resolved by our telescopes, it seems that the proportion of bars is comparable in early galaxies. This is an important finding, and it tells us much about evolutionary processes. The first galaxies had plenty of gas, and therefore their bars must have been short-lived, given the dynamical processes described above. A high rate of accretion of gas would have been necessary to re-form these bars. Therefore, the secular evolution scenario must also have played a central role in the past.

---

[1] The inverse of the Hubble constant, the Hubble time provides an estimate for the age of the universe by assuming that the universe has always been expanding at its current rate. If the universe expanded faster in the past than it does today, then the Hubble time overestimates the age of the universe. If the universe is expanding faster today than it did in the past, then the Hubble time underestimates the true age of the universe.

108 Mysteries of Galaxy Formation

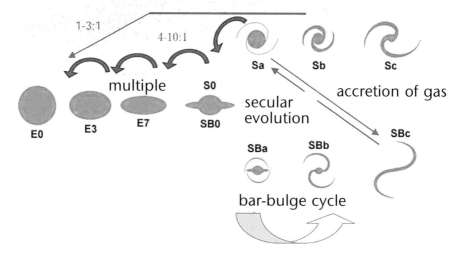

Figure 4.12 The possible scenarios of galactic evolution, combining the hierarchical and the secular evolution theories, are here illustrated in the Hubble Sequence, which classifies galaxies: two branches merge into one, in the manner of a tuning fork. A galaxy can pass from the 'normal' (upper) branch to the 'barred' (lower) branch through gravitational instability in a cool disk (for example, shortly after accretion of external gas). The bar may also be destroyed, by gas flowing into the center. Resonances between stars and the bar may engender the formation of a bulge. The bulge/disk ratio increases and galaxies evolve towards the left-hand side of the diagram. After accretion of gas from cosmic filaments, the disk's mass increases, and evolution then briefly heads towards the right-hand side of the diagram. Finally, interactions and mergers with neighboring galaxies may form more massive spheroidal systems, and the galaxies evolve towards the single (left-hand) branch of the 'tuning fork,' at a rate which depends on the ratio of the masses of galactic neighbors. This rate is rapid if the ratio is between 1:1 and 3:1.

Environmental effects

An important factor in the evolution of galaxies is the density of their environment. In a dense cluster (Figure 4.13), galaxies evolve much more rapidly, maturing into elliptical and lenticular systems, and star formation ceases much earlier on than in 'field' galaxies, i.e. galaxies which do not belong to a cluster. This effect has long been known in the case of local galaxies, from observations of the morphology of galaxies in relatively nearby clusters (e.g. the Virgo, Fornax and Coma clusters); but it is possible to see evolution proceeding even today, by observing clusters of galaxies at greater redshifts. The Butcher-Oemler effect, which has been recognized for about twenty years, involves an ever higher fraction of blue galaxies in clusters as redshift increases. The blue color of these galaxies is due to a high rate of star formation within them.

Scenarios of galaxy formation 109

Figure 4.13 Hubble Space Telescope image of the magnificent Coma Cluster of galaxies, one of the densest known galaxy collections in the universe. Hubble's Advanced Camera for Surveys viewed a large portion of the cluster, spanning several million light-years across (NASA, ESA, and the Hubble Heritage Team (STScI/AURA)). See also PLATE 19 in the color section.

In earlier times, therefore, there existed many more galaxies that were still active within clusters; while today, the majority of galaxies are elliptical and evolved, dominated by old populations of stars and lacking interstellar gas. The morphological segregation of galaxies according to their environment has been known about for some thirty years now. Astronomers have observed that, in rich clusters, nearly 90 percent of the galaxies are elliptical or lenticular (thick disks without gas), while more than 70 percent of 'field' galaxies (i.e. galaxies outside clusters) are spirals. A very lively debate has developed: is this segregation a product of the transformation of galaxies through environmental effects, or are galaxies born with these different morphologies in regions of space containing superdense matter? This is the so-called 'nature or nurture' debate.

The many dynamical effects of the environment are well able to transform galaxies. The high density of galaxies multiplies the possibility of collisions: either fairly slow encounters between galaxies which lead to mergers, and the

110 **Mysteries of Galaxy Formation**

Figure 4.14 Abell 1689, shown in this composite image, is a massive cluster of galaxies located about 2.3 billion light-years away that shows signs of merging activity. Hundred-million-degree gas detected by NASA's Chandra X-ray Observatory surrounds the cluster of galaxies shown in the optical data from the Hubble Space Telescope. The X-ray emission has a smooth appearance, unlike other merging systems such as the Bullet Cluster shown in Figure 6.6. The temperature pattern across Abell 1689 is more complicated, however, possibly requiring multiple structures with different temperatures. The long arcs in the optical image are caused by gravitational lensing of background galaxies by matter in the galaxy cluster, the largest system of such arcs ever found. Further studies of this cluster are needed to explain the lack of agreement between mass estimates based on the X-ray data and on the gravitational lensing. Previous work suggests that filament-like structures of galaxies are located near Abell 1689 along our line-of-sight to this cluster, which may bias mass estimates using gravitational lensing (X-ray: NASA/CXC/MIT/E.-H Peng *et al.*; Optical: NASA/STScI). See also PLATE 20 in the color section.

Scenarios of galaxy formation 111

transformation of spirals into ellipticals, or high-speed, repeated encounters ('galactic harassment'), which encourage secular evolution and transformation towards more evolved morphologies. Moreover, clusters are not only exceptional accumulations of galaxies; they also possess a concentration of hot intergalactic gas, which emits strongly in the X-ray domain (Figure 4.14). Moreover, this hot gas represents the greater part of the visible mass of dense clusters! This gas exerts a dynamical pressure upon fast-moving galaxies within clusters: they experience an intergalactic 'wind,' capable of sweeping away their interstellar medium. Gas swept out of the galaxies, and stripped off by interactions, enriches the intergalactic medium, which becomes ever denser. Collisions not only transform the morphology of galaxies, but 'blow away' their gas: a very effective way of halting star formation.

All these factors encourage a more rapid evolution of galaxies in clusters, compared with 'field' galaxies. The clusters themselves are dynamical structures, forming and changing over time. In the beginning, a cluster as such does not exist. However, smaller structures are present: groups of galaxies of various sizes. It is therefore probable that part of the transformation of galaxies occurs within these groups, which later merge into clusters. For example, compact groups are small entities, containing a few galaxies, with a shallow gravitational 'well.' So the relative velocities of galaxies are not very high compared to their internal rotational velocities, and encounters between galaxies are very likely to result in mergers. Most mergers of spirals to form ellipticals have occurred within these groups; nowadays, relative velocities in clusters are too great (at least five times greater than rotational velocities) to produce galactic mergers.

On the other hand, mergers of groups give rise to clusters very rich in hot intergalactic gas; the greater the mass of the cluster, and the deeper the gravitational 'well,' the hotter the gas becomes. The supply of cool gas to galaxies ceases, arresting star formation. Not only is the interstellar gas of galaxies swept away by the intergalactic wind, but also, the filaments of matter and gas which fed the field galaxies have disappeared. Hot gas can no longer cool and 'nourish' the galaxies. The cooling time of the gas is longer than the age of the universe itself, except at the center of the cluster with its enhanced density, providing a shelter for a flow of cool gas towards the central galaxy. Observations of clusters of galaxies with redshifts of $z = 0.25$ (corresponding to an epoch 4 billion years ago) have allowed us to see this evolution 'directly.' Not only does the number of blue galaxies (where stars are forming) increase as we look back in time, but the fraction of spiral galaxies also increases. It has fallen quite low today, and is now 10 percent! The 'nature and nurture' debate has come a long way; and the answer is not a simple one.

The environment undoubtedly plays a part in subsequent transformations of galaxies (a process continuing today). It has also influenced the fact that initial superdensities, already there in the primordial fluctuations of the universe, are sites where numerous groups of galaxies have been formed, which in their turn form ellipticals and later tend to merge into clusters. The spiral field galaxies which have formed between the groups are then drawn towards the gravitational 'well' of the cluster, and become the blue galaxies in clusters at intermediate redshifts. Before their gas is swept out, and before their surrounding gas is heated, bursts of star formation are triggered within them by the shock of their entry into the hot gas of the cluster, and by interactions with cluster galaxies.

Bimodality between red and blue galaxies

Large-scale sky surveys, such as the Sloan Digital Sky Survey (SDSS) catalogue of galaxies and the 2dF survey (Figures 4.1 and 4.2), have provided evidence, thanks to the statistically useful numbers of objects imaged, of a definite bimodality between two categories of galaxies, distinguished by their colors. This bimodality revisits the former Hubble sequence, with its galaxies classified according to morphology. There exists a sequence of blue galaxies, rich in gas and populated by young stars, and a sequence of red galaxies, with little gas, and dominated by populations of old stars. These two sequences hark back of course to the distinction between spirals and ellipticals (Figure 4.15). This bimodality reveals itself in the luminosity of galaxies (the largest and brightest are the 'red' ones) and in their stellar masses M_{lim}: the boundary between the two categories lies at about 30 billion solar masses, which is the approximate mass of our own Galaxy. The two sequences are also distinguished by their rates of star formation (high in the blues, and low in the reds), the size of their central bulges compared with their disks (large in reds, and smaller in blues), and their environments (rich in reds, and poor in blues). Low-mass (blue) galaxies are usually 'field' spirals within which star formation is occurring. More massive (red) galaxies are ellipticals with aged populations, and are found in clusters. Observations of high-redshift galaxies show that a large fraction of the red sequence is already in place at z = 1-2, and that there are massive star-forming galaxies earlier in the universe at z = 2-4.

How can we explain the paradoxical phenomenon that the only currently active galaxies are the smallest? Does this not contradict the hierarchical scenario, which holds that small galaxies formed first, early on in the universe, and then merged progressively to form today's larger galaxies? Some astronomers are minded to abandon the hierarchical scenario in favor of monolithic collapse, with massive elliptical galaxies forming all at once, at the beginning of the universe. However, this is not really necessary: it is always possible (and probable) that massive ellipticals are the product of mergers of spirals which were themselves formed early on, in rich environments corresponding to the superdensities of the young universe.

On closer examination, the paradoxical observation of active massive galaxies early on in the universe and active small galaxies in today's universe is not so surprising, especially if we take into account environmental effects and bear in mind the important role of the secular evolution of galaxies through accretion of external gas. In fact, the rate at which evolution takes place can be very variable,

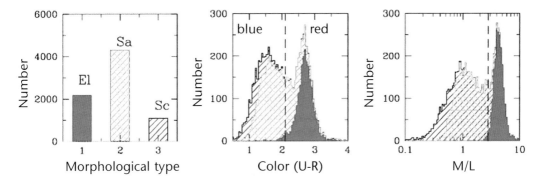

Figure 4.15 Bimodality between 'blue' and 'red' galaxy sequences. **Left:** galaxies classified into ellipticals (red), Sa (green) and Sc (blue). The histogram of the same galaxies as a function of their colour (U-R, or the difference between their ultraviolet and red magnitudes) clearly shows two peaks rather than a continuous distribution. The right-hand (red) peak essentially represents galaxies which are elliptical and well evolved (including a few Sa galaxies), while the left-hand peak represents exclusively spirals. **Right:** the histogram of the same galaxies as a function of the mass/ luminosity ratio of their stellar component also shows two distinct peaks: again, the right-hand peak consists largely of ellipticals which have old, dim stars per unit of mass (after Driver *et al.*, 2006).

depending on the environment. In initially superdense regions, things happened much more quickly: the first galaxies, even if they were dwarfs to begin with, evolved over periods of between one and two billion years; it is not possible to verify this, except in simulations, because the sensitivity of our instruments is such that they can detect only very massive objects in the very distant universe. In the main, these objects were formed in regions where clusters of galaxies later arose, and as described above, clusters of galaxies are very efficient at stopping evolution and suppressing star formation. In the 'field,' on the contrary, evolution is delayed, which is why spirals and dwarf galaxies still exhibit star formation, and evolution has not yet been curtailed by the environment.

One question remains: why should there be such a marked bimodality in the colors of the two sequences of galaxies, instead of a continuous distribution? Several hypotheses have been developed, invoking, for example, the masses of dark-matter halos, and self-regulation phenomena due to supernovae or active nuclei. One hypothesis states that the limiting mass of stars M_{lim} plays a less important role than the surface density of the stars in galactic disks. The sequence of (massive) red galaxies is also one of highly concentrated galaxies, with high surface density of stars (or high surface brightness), as shown in Figure 4.16. What distinguishes the two sequences is the increasingly important presence of a bulge or spheroid in the 'massive' sequence. In fact, on one side, the galaxies are dominated by their disk, and on the other, by their bulge. The properties of stellar formation are therefore associated with these two universal components.

114 Mysteries of Galaxy Formation

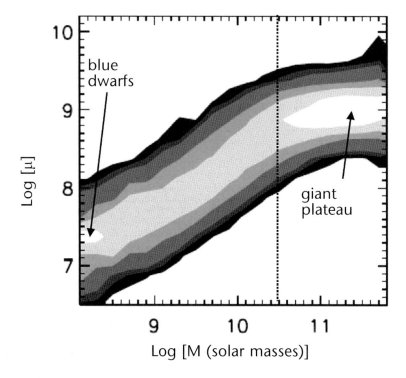

Figure 4.16 The most massive galaxies have high (and almost constant) surface density, while dwarf galaxies have low surface density. In this log-log diagram, the vertical axis represents the mean surface density µ of the stars of a galaxy as a function of its total stellar mass (horizontal axis). The contours quantify the number of galaxies populating the various regions of the diagram, each contour representing a factor of 2 in numbers of galaxies. On the right, there appears a limiting surface density for massive 'red' galaxies. The plateau for density µ ceases at stellar mass limit $M_{lim} = 3.10^{10}$ $M_{\odot}$, shown on the diagram by a vertical dotted line. Beyond this, surface density decreases for the sequence of 'blue' galaxies (after Kauffmann et al., 2003).

The size of galaxies increases with mass, but in the case of massive galaxies, above M_{lim}, this increase is more rapid. The efficiency of star formation in the past is greater for massive galaxies, which have transformed their baryons more rapidly into stars. As a consequence, the fraction of dark matter is proportionally greater in small galaxies. The essential underlying phenomenon of these ratios is very probably connected with self-regulation by supernovae. In less massive galaxies, the efficiency of star formation is limited early on, as stellar winds and supernovae can easily expel gas from a low-mass system: its escape velocity is relatively low. When the escape velocity increases and becomes greater than the impetus velocity of supernovae, which is unchanging, gas is no longer expelled from the galaxy and star formation can continue unchecked.

Scenarios of galaxy formation 115

The limiting size of galaxies, at the boundary of these two regimes, is not well known, since it depends on dark matter and its concentration; we can surmise that the critical mass of the dark halo intervenes for a limiting mass of stars equal to M_{lim}. Introducing regulating phenomena due to supernovae gives a clue as to why star formation is delayed in small systems, from which the gas is often expelled. Also, star formation may cease in massive systems, even if they have come together only relatively recently: in these systems, dark halos have been massive since the beginning, and have retained the gas, which has been very efficiently forming stars. Today, intermediate-mass systems merge, but without gas – it has already been transformed into stars. Massive galaxies are therefore recent ones, and are not as old as their stars.

A second approach considers that all the while the dark halo is not deep enough to heat, through shocks, the gas feeding into the galaxies, star formation can continue, as happens in small blue galaxies. When the mass exceeds a critical value, the gas is heated by shocks. Its cooling time becomes greater than the age of the universe and the lack of cool gas halts star formation. This approach suggests that the limiting mass is that mass which forms clusters of galaxies. This argument is based essentially on baryonic processes which control the formation of stars. It has long been known that the physical processes involving the heating and cooling of gas are fundamental in explaining the characteristic scales of galaxies. However, dark-matter halos obey only the laws of gravity, and have no characteristic scale. Their properties are similar over a large range of scales. Since the cooling time of the gas is an increasing function of the mass of the structure in question, the upper limiting mass that a galaxy can have is established by equating the characteristic cooling time to its characteristic collapse time, or to the age of the universe.

Finally, there is a third approach: this involves the dark halos themselves. When we consider Figure 4.4, and the definition of the time t_f for the formation of a halo of a given mass as the time during which a progenitor halo has formed containing at least half the mass of the final halo, it is certain that the largest halos form much later than less massive ones. On the other hand, if we take into account all the progenitors capable of forming stars, we must consider a minimum mass for sub-halos (let us call it M_{min}) for the gas to be able to cool and condense into stars. In this context we have to add together the 'useful' mass of all the progenitors which have contributed at a given time t to star formation. Since the value of M_{min} is universal, and not proportional to the mass of the final halo in question, this value will be almost negligible for massive halos. However, it will be a limiting element in the case of weak halos. In a way, massive halos already have all their progenitors practically at 'useful mass,' allowing star formation early on in the universe; but weak halos had to await a more recent epoch. In this sense, massive halos (even if they are not yet completely assembled today) have already brought together all their stars at times corresponding to high redshifts, which fits the observations.

Feedback between supermassive black holes and star formation

Another important phenomenon has to be taken into account in the galaxy formation scenario. As we have seen in the previous chapter, there is a supermassive black hole growing in the center of every galaxy, and its growth is proportional to the growth of the bulge mass, related to star formation. This concomitant growth must be based on a regulating mechanism. Certainly, when gas is accreted by a galaxy, the gas is available both for star formation and to fuel the black hole in the center. In massive galaxies, the galaxy nucleus can then be active enough to expel significant amounts of ionized plasma, in jets propagating far away in the galaxy halo. The associated energetic phenomena can heat the inflowing gas, and stop star formation. Since the black-hole mass is proportional to the bulge mass, this negative feedback process is more efficient in massive galaxies, where the star formation feedback is less efficient. The active nucleus phases, corresponding to gas accreted by the black hole, are short in the life of a galaxy, typically of the order of a few per cent, given the rarity of the phenomenon in the observations. However, this can provide a sufficient heating mechanism to explain the suppression of star formation in massive galaxies.

Dwarf elliptical or dwarf spheroidal galaxies

One class of galaxies seems to pose a problem in this scheme of evolution. These are the low-mass galaxies which have already lost all their interstellar gas and are no longer evolving. How can we reconcile them with the classification based on current theories? These theories provide for:

- small galaxies, already endowed with a large fraction of gas and actively forming stars, with a low bulge/disk ratio (blue galaxies sequence);
- more massive galaxies, low in gas, with star formation completely or almost over, dominated by their concentrated central spheroid (red galaxies sequence).

Where do these galaxies, low in mass but spheroidal, lacking gas and young stars, fit in? They can be of two types:

- their stars may be very concentrated (compact dwarf ellipticals);
- or very diffuse (dwarf spheroidals).

These dwarf galaxies, containing little gas, are observed in very particular contexts. Compact dwarf galaxies especially are relatively rare, and are always observed as satellites of larger galaxies. A typical example is the galaxy M32 (Figure 1.16), a close companion of the Andromeda Galaxy M31.

There exists also a category known as ultra-compact dwarfs (UCDs). These are always observed in clusters of galaxies. Discovered in the Fornax cluster, they also exist in the Virgo cluster. Judging by their velocity dispersions, they do not seem to sit in their own halo dark matter, but belong to the larger halo of the companion or the galaxy cluster. Several hypotheses have been put forward as to their origin: some of these galaxies could be spirals whose disks have been destroyed or stripped by tidal interactions with neighboring galaxies; the gas might have been swept away by intergalactic winds in clusters. UCDs resemble globular star clusters, but they are a hundred times brighter. In many of their properties, they represent a transition between globular star clusters and galactic nuclei. Some of them are rather red in color, while others are blue; but in all cases they are dominated by a population of old stars. Over and above the hypothesis that they represent the nuclei of larger galaxies whose disks have been destroyed, it has been suggested that they could be products of the agglomeration of several giant star clusters that were formed when galaxies merged. Once the mergers were completed, these agglomerations would have been released into intergalactic space. Unlike globular star clusters, UCDs are not observed near parent galaxies, but orbit at the cores of rich galaxy clusters.

Dwarf spheroidal (or low-surface brightness) galaxies are more numerous. In our local group, about thirty have been identified, accompanying the Milky Way (Figure 4.17) and the Andromeda Galaxy. They have mostly been found in recent times, because they are very diffuse systems, and very difficult to detect against the background stars of our Galaxy. These dwarf galaxies could be more easily located within the bimodal classification described above, if they were richer in gas. It is tempting to see them as objects which have lost their gas through multiple interactions, or because of dynamical pressure while traversing other gaseous mediums at high velocity. They are systems with high mass/luminosity ratios, and therefore seem to possess a massive dark matter halo. However, this ratio increases in the proximity of a giant neighboring galaxy, which suggests tidal effects which could artificially increase the apparent dark matter, as deduced from kinematics.

In effect, the stars in the tidal debris would no longer be linked to the dwarf galaxy, and could exhibit very high velocities. The history of star formation in these systems is well known, as, being relatively close, they have been resolved into individual stars. Although they are dominated by populations of old stars, some show evidence of more recent episodes involving the formation of stars (of intermediate age). However, no stars are forming in them today. Dynamical models suggest that repeated tidal interaction with the neighboring giant galaxy could explain the transformation of spiral galaxies into dwarfs, and the origin of dwarf spheroidals.

The geometrical disposition of dwarf satellite galaxies does not appear to be completely isotropic in the vicinity of giant galaxies. For example, as Figure 4.18 shows, they seem to lie in a plane which is almost perpendicular to the plane of the Milky Way. The same is true in the case of the Andromeda Galaxy, where the satellite galaxies seem to orbit in a plane rather inclined to that of the disk of the main galaxy. Does this disposition tell us anything about their origins? It could be that the dwarf galaxies are themselves formed from the debris left by the tidal interactions which gave rise to the giant galaxies of the Local Group. This would account for their non-random orientation: but there are too many of them. A

118 Mysteries of Galaxy Formation

Figure 4.17 One of the newly discovered low-luminosity dwarf spheroidal galaxies, accompanying the Milky Way, from the Sloan Digital Sky Survey, This galaxy, named Leo T, apparently lies at a distance of 420 kpc, and shows evidence for an ancient low-metallicity population, as well as for fairly recent star formation (M.J. Irwin and the Sloan Digital Sky Survey (SDSS-II)). See also PLATE 21 in the color section.

more likely explanation is that the satellite galaxies are oriented according to cosmic filaments, where all galaxies form in numerical simulations, and whose existence seems confirmed by the major galaxy surveys (Figures 4.1 and 4.2). The orientation of the planes of giant galaxies almost at right angles to filaments is indeed reproduced by these simulations. This orientation is a 'memory' of the original angular momentum of the disks of galaxies in rotation. However, after several mergers of galaxies along filaments, groups or giant elliptical galaxies are oriented parallel to the filaments.

PLATE 1 = Figure 1.1 The Andromeda Galaxy (M31) is the closest large spiral galaxy to our own Milky Way. Located 2.5 million light-years away one can easily locate it with the naked eye in the constellation of Andromeda on clear, moonless nights. This overlapping three frame mosaic shows the galaxy and its two small satellites M32 (above center) and NGC 205 (below left of center). The dynamic range of the combined image has been compressed significantly to show the inner and outer regions of the galaxy (Adam Block, NOAO, AURA, NSF).

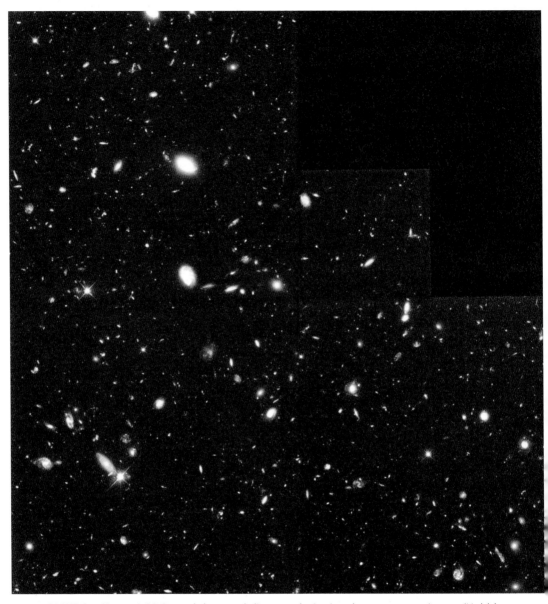

PLATE 2 = Figure 1.10 Several thousand distant galaxies in a long-exposure image (Hubble Deep Field North). A true-color image from the Hubble Space Telescope of a small area in the northern hemisphere of the sky. This region was observed in 1996, and this 10-day exposure was built up from 342 separate images. Although the region of sky shown is only 2.5 minutes of arc across, more than 3,000 galaxies can be identified, thanks to the high sensitivity and good quality of the image (the resolution being 0.1 second of arc). It was published as soon as it was obtained, so that ground-based spectroscopic research could be concentrated on this region, chosen because there are so few foreground objects in our own Galaxy (R. Williams (STScI), the Hubble Deep Field Team and NASA).

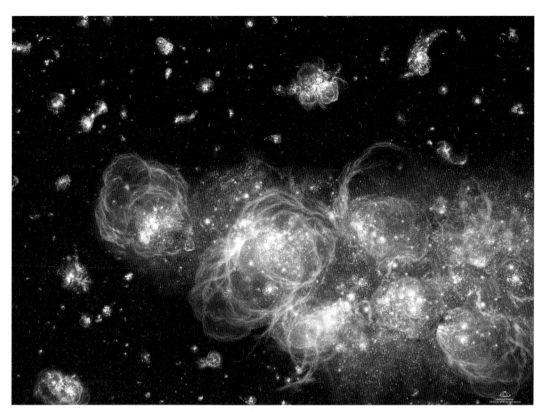

PLATE 3 = Figure 1.17 An artist's impression of how the very early universe (less than one billion years old) might have looked when it went through the voracious onset of star formation. The most massive of these Population III stars self-detonated as supernovae, which exploded across the sky like a string of firecrackers (Adolf Schaller for STScI).

PLATE 4 = Figure 1.20 A near-infrared view of our Galaxy. This panorama encompasses the entire sky as seen by the Two Micron All-Sky Survey (2MASS). The image is centered on the core of the Milky Way, toward the constellation of Sagittarius, and reveals that our Galaxy has a very small bulge (normal for a late-type Sbc galaxy), which has the shape of a box or peanut (University of Massachusetts, IR Processing & Analysis Center, CalTech, NASA).

PLATE 5 = Figure 2.7 This million-second-long exposure, called the Hubble Ultra Deep Field (HUDF), reveals the first galaxies to emerge from the so-called 'dark ages,' the time shortly after the Big Bang when the first stars reheated the cold, dark universe. This view is actually two separate images taken by Hubble's Advanced Camera for Surveys (ACS) and the Near Infrared Camera and Multi-object Spectrometer (NICMOS) (NASA, ESA, S. Beckwith (STScI) and the HUDF Team).

PLATE 6 = Figure 2.10 Simulations of clumpy galaxies. The first galaxies are constituted almost entirely of gas that settles in a disk, through dissipation. The disk is dynamically cold, and highly unstable against gravitational collapse into clumps. Progressively, clumps are driven into the center by dynamical friction, and form a bulge, which stabilizes the disk (from Bournaud et al., 2007).

PLATE 7 = Figure 3.1. Accretion disk around a black hole in a binary star system, such as Cygnus X-1. This artist's impression shows an 'ordinary' star (left) whose gaseous envelope is so drawn out that it has reached the Roche lobe, the boundary between the respective spheres of attraction of two stellar companions. The matter lost by the star is accreted onto its companion, which has already collapsed to become a black hole (right) (NASA/STScI and ESA).

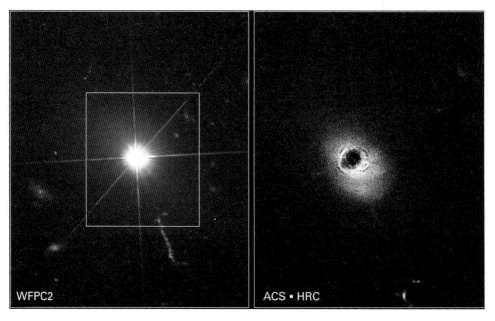

PLATE 8 = Figure 3.3 **Left:** this image of the quasar 3C273 in the constellation of Virgo, acquired by the Hubble Space Telescope's Wide Field Camera 2, shows the brilliant quasar but little else (NASA and J. Bahcall (IAS)). **Right:** once the blinding light from the brilliant central quasar is blocked by the coronagraph of Hubble's Advanced Camera for Surveys (ACS), the host galaxy pops into view (NASA, A. Martel (JHU), H. Ford (JHU), M. Clampin (STScI), G. Hartig (STScI), G. Illingworth (UCO/Lick Observatory), the ACS Science Team and ESA).

PLATE 9 = Figure 3.4 Hubble Space Telescope face-on view of the small spiral galaxy NGC 7742. This is a Seyfert 2 active galaxy, a type of galaxy that is probably powered by a black hole residing in its core. The core of NGC 7742 is the large circular 'yolk' in the center of the image (The Hubble Heritage Team (STScI/AURA) and NASA).

PLATE 10 = Figure 3.6 Streaming out from the center of the galaxy M87 is a black-hole-powered jet of electrons and other sub-atomic particles traveling at nearly the speed of light. Lying at the center of M87 is a supermassive black hole, with a mass equivalent to over 2 billion times the mass of our Sun. The jet originates in the disk of superheated gas swirling around this black hole and is propelled and concentrated by the intense, twisted magnetic fields trapped within this plasma. (NASA and the Hubble Heritage Team (STScI/AURA)).

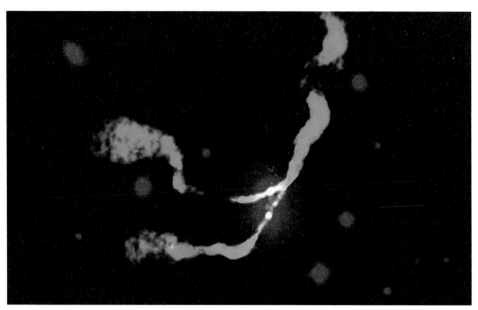

PLATE 11 = Figure 3.9 The radio source 3C 75 consists of the synchrotron radio emission of the radio jets issuing from two galaxies at the center of the cluster Abell 400. The two pairs of jets originate in the black holes in the nuclei of the two central galaxies. These galaxies are moving at high speed through the cluster, which is filled with very hot gas emitting X-rays (NRAO).

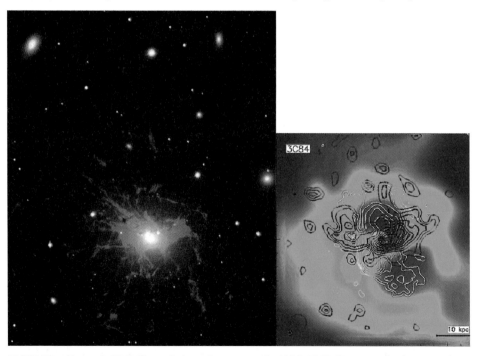

PLATE 12 = Figure 3.13 Self-regulation phenomena in NGC 1275 (Perseus A), the central galaxy of the Perseus cluster. **Left:** a deep hydrogen-alpha image of NGC 1275, taken by the WIYN 3.5-m telescope at Kitt Peak National Observatory. The filaments emanating from this galaxy are probably the result of an interaction between the black hole in the center of the galaxy and the intracluster medium surrounding it (C. Conselice/Caltech and WIYN/NOAO/AURA/NSF). **Right:** an image of the X-ray emission from the very hot gas in the cluster (Chandra X-ray Observatory, Fabian *et al.*, 2000). The contours shown in white represent the emission from radio jets (Pedlar *et al.*, 1999).

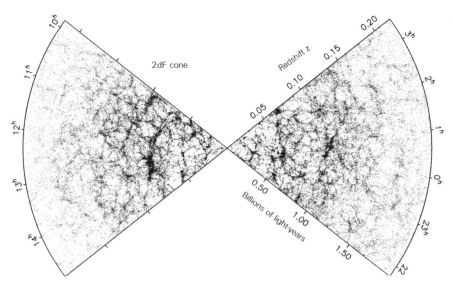

PLATE 13 = Figure 4.1 Three-dimensional survey of the large-scale structures of the universe. In this 2dF (2-degree Field) survey, in addition to the two dimensions of the plane of the sky, there is a third dimension obtained by measuring the velocities or redshifts of 221,000 galaxies. The 2dF survey, by the Anglo-Australian Telescope (AAT), is based on areas of the sky measuring two square degrees. Each dot on this chart is a galaxy, whose distance in billions of light years is inferred from its radial velocity. Galaxies are distributed in clusters and superclusters, along a filamentary structure around bubble-like voids, the whole reminiscent of a cobweb.

PLATE 14 = Figure 4.3 Detailed computer simulations are a powerful tool in the study of how the galaxies we see in the universe today formed out of the small fluctuations in matter density that were present in the primordial universe. In this single frame, a slice through a small part of the simulated universe, the large-scale structure in the dark matter is seen (Max-Planck-Institut für Astrophysik and the Virgo consortium).

PLATE 15 = Figure 4.5 This image from the Hubble Space Telescope's Advanced Camera for Surveys shows the Antennae galaxies – seemingly a violent clash between a pair of once isolated galaxies, but in reality a fertile marriage. As the two galaxies interact, billions of stars are born, mostly in groups and clusters of stars. The brightest and most compact of these are called super star clusters. The two spiral galaxies started to fuse together a few hundred million years ago making the Antennae galaxies the nearest and youngest example of a pair of colliding galaxies. Nearly half of the faint objects in the Antennae are young clusters containing tens of thousands of stars (NASA, ESA, the Hubble Heritage Team (STScI/AURA), and the ESA/Hubble Collaboration, with thanks to B. Whitmore (STScI) and James Long (ESA/Hubble)).

PLATE 16 = Figure 4.6 A spectacular pair of interacting galaxies captured by the Hubble Space Telescope's Advanced Camera for Surveys. Located 300 million light-years away in the constellation Coma Berenices, the colliding galaxies have been nicknamed 'The Mice' because of the long tails of stars and gas emanating from each galaxy (NASA, H. Ford (JHU), G. Illingworth (UCSC/LO), M.Clampin (STScI), G. Hartig (STScI), the ACS Science Team, and ESA).

PLATE 17 = Figure 4.8 M80 (NGC 6093), one of the densest of the 147 known globular star clusters in the Milky Way galaxy, captured by the Hubble Space Telescope's Wide Field and Planetary Camera 2. Located about 28,000 light-years from Earth, M80 contains hundreds of thousands of stars, all held together by their mutual gravitational attraction (NASA, the Hubble Heritage Team (STScI/AURA), with thanks to M. Shara and D. Zurek (AMNH), and F. Ferraro (ESO)).

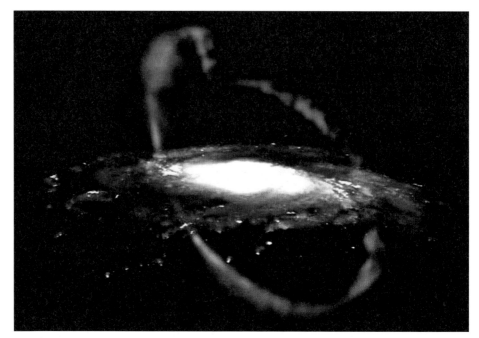

PLATE 18 = Figure 4.9 Artist's impression of our Galaxy, the Milky Way, surrounded by a stream of stars. Numerous stellar streams and tidal debris have recently been discovered around and in the halo of the Milky Way. This artist's impression shows a star stream recently 'peeled' off the Sagittarius dwarf galaxy. This stream will long pursue the dwarf galaxy's former orbit around the Milky Way.

PLATE 19 = Figure 4.13 Hubble Space Telescope image of the magnificent Coma Cluster of galaxies, one of the densest known galaxy collections in the universe. Hubble's Advanced Camera for Surveys viewed a large portion of the cluster, spanning several million light-years across (NASA, ESA, and the Hubble Heritage Team (STScI/AURA)).

PLATE 20 = Figure 4.14 Abell 1689, shown in this composite image, is a massive cluster of galaxies located about 2.3 billion light-years away that shows signs of merging activity. Hundred-million-degree gas detected by NASA's Chandra X-ray Observatory surrounds the cluster of galaxies shown in the optical data from the Hubble Space Telescope. The X-ray emission has a smooth appearance, unlike other merging systems such as the Bullet Cluster shown in PLATE 23 = Figure 6.6. The long arcs in the optical image are caused by gravitational lensing of background galaxies by matter in the galaxy cluster, the largest system of such arcs ever found (X-ray: NASA/CXC/MIT/E.-H Peng *et al.*; Optical: NASA/STScI).

PLATE 21 = Figure 4.17 One of the newly discovered low-luminosity dwarf spheroidal galaxies, accompanying the Milky Way, from the Sloan Digital Sky Survey, This galaxy, named Leo T, apparently lies at a distance of 420 kpc, and shows evidence for an ancient low-metallicity population, as well as for fairly recent star formation (M.J. Irwin and the Sloan Digital Sky Survey (SDSS-II)).

PLATE 22 = Figure 6.2 Starburst activity in the 'Cigar galaxy' M82, and a manifestation of self-regulating energetic phenomena. **Left:** visible-light image of the galaxy, seen edge-on, shows only a bar of light, mottled with dust lanes, against a dark patch of space. **Right:** mid-infrared image at wavelengths between 3.6 and 8 microns, from NASA's Spitzer Space Telescope. Old stars can be seen in the component seen edge-on, and dust is being ejected at right angles to the plane, in a bipolar flow, blown out into space by the galaxy's hot stars (Visible: NOAO; Infrared: NASA/JPL-Caltech/C. Engelbracht (University of Arizona)).

PLATE 23 = Figure 6.6 Composite image of a collision between two clusters of galaxies. Superposition of X-ray emissions from hot gas and the projected mass in the galaxy cluster 1E 0657-56, familiarly known as the 'Bullet Cluster' (X-ray: NASA/CXC/CfA/M.Markevitch *et al.*; Lensing Map: NASA/STScI, ESO WFI, Magellan/U.Arizona/D.Clowe *et al.*; Optical: NASA/STScI, Magellan/U.Arizona/D.Clowe *et al.*). The projected X-ray gas and masses are also superimposed on the optical image showing the individual galaxies. The different behavior during the collision of the hot gas and the stellar masses, including dark matter, allows us to separate the three components and test the models (after Clowe *et al.*, 2006).

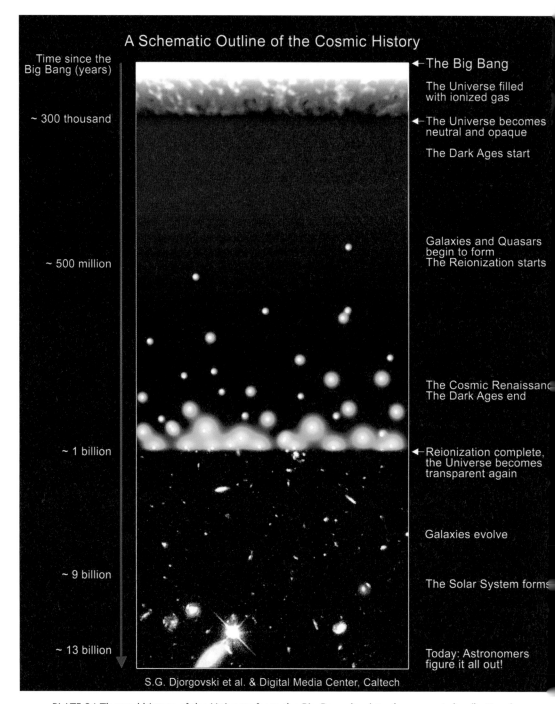

PLATE 24 Thermal history of the Universe from the Big Bang (top) to the present day (bottom) 13.7 billion years later (S.G. Djorgovski et al., Caltech and the Caltech Digital Media Center).

Scenarios of galaxy formation 119

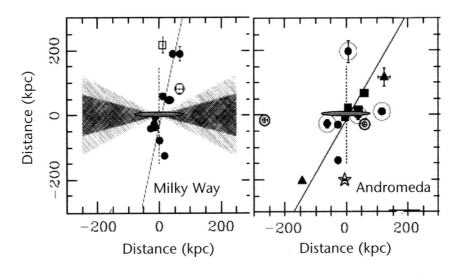

Figure 4.18 The distribution of the dwarf satellite galaxies of the Milky Way (left) and the Andromeda Galaxy (right). The giant galaxies are seen edge-on, and are shown here as horizontal flat ellipses. The grey areas in the left-hand diagram represent areas of sky obscured by the dust of the Milky Way. The symbols represent the main satellite galaxies, all lying along a plane (also edge-on in the diagrams). These planes may represent the local filament of matter with which the giant galaxies are associated. The scale is in kiloparsecs (after Metz et al., 2006).

It may seem that the Milky Way is well endowed with dwarf spheroidal galaxies (there are a dozen in its vicinity), but in fact this number is considered to be very low in the context of the theory involving CDM and the hierarchical formation of structures. The simulations all predict the existence of several hundred small dark halos of this type which could have given rise to dwarf galaxies around our Galaxy, as will be seen in greater detail in the next chapter. It seems therefore that, on the scale of a galaxy such as our own, our vision of the formation of galaxies is still quite far from coinciding with the observations.

5 The problem of dark matter

Dark matter plays a crucial role in the formation of galaxies. The very nature of this matter and its interaction with other particles determines the proportion of large and small structures able to form at the beginning of the universe.

The CDM (Cold Dark Matter) model has been highly successful in explaining the distribution of large-scale structures. However, on the scale of galaxies, difficulties have arisen.

What is the relationship between dark matter and visible matter? The two kinds of matter evolve together within large structures, but, in galaxies, this parallel development no longer applies, giving way to an imbalance, a bias. In an attempt better to understand these differing evolutions, the scaling relations and the physical properties of galaxies are observed as a function of redshift.

If it proves impossible to resolve the current problems with the standard cosmological model, through better knowledge of the dynamics of gas and the formation of stars, will we have to resort to other, more drastic hypotheses, such as changing the law of gravity?

The large-scale structure of the universe: the success of the CDM (Cold Dark Matter) model

As we saw in Chapter 1, the formation of structures through gravitational collapse was very slow when the universe was young, due to expansion, which moved the various masses away from each other. The presence of non-baryonic dark matter allowed structures to start forming earlier, giving more time for collapse: thus, the presence of galaxies today is explained. Moreover, this dark matter is not only needed to explain the 'missing mass' in galaxies and clusters; it has a profound role to play.

Quantitative measurements of the cosmic background radiation (the vestigial 'echo' of the Big Bang) have led to the determination of the amplitude of primordial density fluctuations, at the time of the last scattering of photons when baryonic matter recombined, to be transformed from plasma into neutral atomic hydrogen about 380,000 years after the Big Bang. These fluctuations were so small relative to density that they could not have given rise to the galaxies of today, if matter consisted only of baryons. Baryons were ionized during the first few hundred thousand years after the Big Bang, remaining coupled with photons, whose pressure prevented structures from collapsing. The baryons could take part in collapse only after the recombination of ions into neutral gas, which occurred 380,000 years after the Big Bang.

It is of absolute necessity that these fluctuations should develop earlier, in a medium that does not interact with either photons or any other components of the universe, except through the forces of gravity. This medium has been named 'dark matter.' The fluctuations were able to develop from the time of the equivalence of radiation and matter, from the moment when the density of matter in the universe began to exceed that of the photons (corresponding to redshift $z \sim 3,200$), 70,000 years after the Big Bang. The fluctuations were then able to be multiplied by a factor of at least three, in comparison with fluctuations of the baryons. The distribution as a function of scale (or of mass) of these fluctuations is predicted by the theory of inflation (a period of very rapid expansion just after the Big Bang), which to date has shown itself to be compatible with observations.

This distribution is represented by its power spectrum $P(k)$ (Figure 5.1), as a function of spatial frequency k, which is the inverse of the λ scale considered ($k = 2\pi\lambda$). This function $P(k)$ represents, in a way, the amplitude of the fluctuations on this scale. The spectrum predicted by inflation is indeed observed by the COBE and WMAP satellites, on a very large scale (with low values of k). In the small-scale spectrum, the distribution is inverted. The pressure of the photons now inhibits the growth of fluctuations in density. The fluctuations are not affected all the while their size is greater than that of the horizon. The latter is very small at the beginning of the universe, in proportion to its age. Once structures enter the horizon, their growth ceases, at least before the epoch of equivalence between matter and radiation. The lack of growth on small scales, as compared with large, causes the power spectrum $P(k)$ to tend towards high values of k (or small scales).

This distribution in the amplitude of structures as a function of scale, predicted by theory and confirmed by cosmological computer simulations, corresponds remarkably well to observations made on different scales by various means (Figure 5.1). From the largest scales to the smallest, all the observations (the cosmic microwave background, the great galaxy surveys, the distribution of clusters of galaxies, and the network of cosmic filaments traced out by the absorption of intergalactic gas) concur to reproduce the power spectrum of the fluctuations $P(k)$ as a function of spatial frequency k (the inverse of the scales). This agreement represents a success for the CDM (cold dark matter) model, associated with the cosmological constant Λ.

If the dark matter had been dominated by a 'hot' component (HDM), composed of still-relativistic particles as it decoupled thermally from the cosmic background, those relativistic particles could have prevented the growth of the fluctuations not only on the small scale, but also on the medium scale, and the spectrum would have been deformed. In particular, there would have been far fewer small-scale structures than other kinds, and far fewer galaxies formed.

The problem of dark matter 123

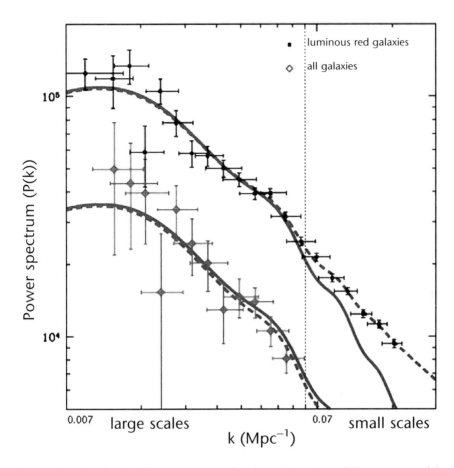

Figure 5.1 Amplitude of fluctuations as a function of size. One of the successes of the CDM model is the prediction of the distribution in scales of the amplitudes of fluctuations, which have been observed in several ways (through studies of the anisotropies in the cosmic microwave background, surveys of galaxies such as the Sloan Digital Sky Survey, and of the abundance of clusters). The diagram shows the power spectrum of the fluctuations P(k) as a function of spatial frequency k (the inverse of the scales). The CDM prediction (+ dark energy) is the curve below the dots representing the observations. The unbroken line is the linear theory of gravitational collapse, valid on large scales, and the dashed line takes account of the non-linear collapse of structures, which soon begins to be significant on small scales. The two samples of galaxies correspond to: the total number of galaxies; and the luminous red galaxies, which are about three times more likely to correlate than the ensemble of galaxies. The latter red galaxies, which are more luminous, are detectable at greater distances, so the volume here considered is much larger, and this accounts for the lesser dispersion. It is well known (see Chapter 4) that luminous red galaxies are generally ellipticals, which are found in densely populated environments (groups or clusters); this explains their higher rate of spatial correlation and their higher P(k) (after Tegmark et al., 2006).

124 Mysteries of Galaxy Formation

Baryonic oscillations: another success for the CDM model

Moreover, modern galaxy surveys have managed to detect oscillations in the spatial distribution of galaxies, or the power spectrum P(k). These 'wrinkles' in the spatial distribution of galaxies are vestiges of the time when the baryons oscillated with photons, before the recombination era. At that epoch, the ionized baryons constituted a plasma, closely coupled to photons, and were unable to collapse under their own gravity, but possible overdensities were in equilibrium with pressure, and participated in acoustic waves along with photons. The waves propagated at the speed of sound from the 'instant zero' of the Big Bang. The 'wrinkles' attained the form of a sphere of radius equal to the sound horizon, when the recombination occurred and the baryons decoupled from the photons. The baryonic wrinkles remained 'frozen' and the galaxies formed, preferentially at the sites of the overdensities.

The size of the 'sound' horizon at the time of the recombination was about 500 million light years: it is thus at this scale, or at the corresponding spatial frequency, that we find the first wrinkle in the power spectrum. Of course, this signature is a very weak one, but it has been detected, and this is a first crucial step in the determination of the whole spectrum of oscillations which will reveal more to us about the nature of dark matter, and the variation with time of the parameters of dark energy. To date, only the first wrinkle has been detected, and for very local galaxies. In the future, it will be possible to detect various oscillations, the equivalent of baryons in the anisotropies already detected in the cosmic microwave background by the WMAP satellite (Figure 5.2). Given that

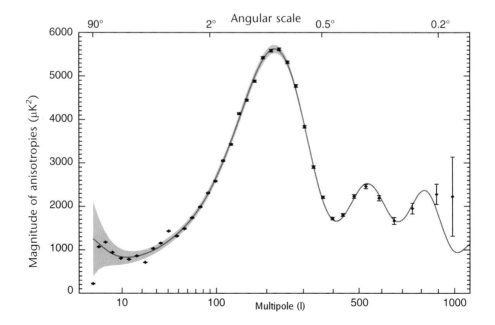

The problem of dark matter 125

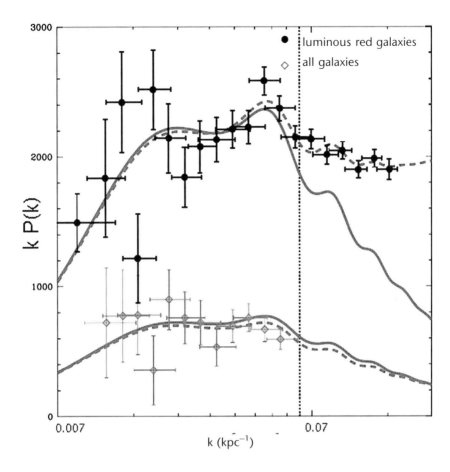

Figure 5.2 Sound waves (or acoustic oscillations) in the cosmic microwave background and in the distribution of galaxies (baryons). **Opposite:** anisotropies in the cosmic microwave background as a function of spatial frequency (WMAP image). Observations are shown with error bars due to noise. The curve is the fit of the CDM theory to the observations (symbols), including dark energy. The shaded area represents cosmic variance, i.e. the uncertainty resulting from the low number of very large structures measured within our finite horizon. **Above:** corresponding acoustic oscillations from baryons, obtained by multiplying the power spectrum P(k) by k, in order to reveal the wrinkles in the curve (see Figure 5.1). Catalogues of galaxies indeed show the first oscillation, and the power spectrum proceeds faithfully along the non-linear curve of the models (after Tegmark et al., 2006).

the intrinsic size of these 'wrinkles' is known, measuring them at various epochs in the age of the universe will enable us to measure the expansion as a function of time. These oscillations act as 'standard rulers' for measuring the evolution through time of the characteristic scale of the universe.

The first wrinkle was detected simultaneously in the spatial direction along the line of sight, where distances are defined by Hubble's law of expansion, and on the plane of the sky by direct measurement. Since these scales must be strictly identical, comparing them permits a more accurate measurement of the Hubble constant. It has already been possible to carry out this measurement at the surface of last scattering, 380,000 years after the Big Bang, by means of the cosmic microwave background and WMAP. However, this measurement involves only one epoch, when the dark-energy component was not dominant. It will be crucial to perform measurements at different epochs with acoustic baryonic oscillations, thereby measuring the geometry of the universe and its contents. In the future, spectroscopy involving millions of galaxies will achieve this goal. This method is complementary to that using Type Ia supernovae as standard candles to measure the geometry of the universe.

Does visible matter follow dark matter? The bias

In the standard CDM model, baryons fell, after the recombination, into potential wells that had already formed in the dark matter. But the physics of the two components were not the same, and it is logical to imagine that there was no perfect coincidence of visible and dark matter, at least on small scales. The processes of dissipation, star formation, etc. made it totally improbable that there could be a coincidence of visible matter and total mass. We know, for example, that there exists a clear segregation of morphological types in clusters, and that massive galaxies (usually red ellipticals) are much more gregarious than other types: each color filter offers a different vision of these large structures.

This difference between visible mass and dark matter is known as the bias (b), representing the ratio of the two densities. The large surveys of slices in the universe have shown that this quantity is equal to one, i.e. that there is no large-scale bias, and that visible matter is a good tracer of total mass. However, because of small-scale gaseous processes, this is not true at the level of galaxies. The bias can also be measured using the gravitational lensing technique, on various scales. Clusters of galaxies deform, with their mass, light rays from background galaxies, and mapping these makes it possible to measure mass distribution in the 'lens,' i.e. the cluster. This method has shown that dark matter seems more concentrated than visible matter in clusters of galaxies. This observational fact sets constraints upon the nature of dark matter.

Farther out from the cluster of galaxies, the phenomenon of gravitational lensing due to all the matter projected onto the plane of the sky produces very slight distortions of the images of the background galaxies. As opposed to the phenomenon of strong lensing, this 'shearing' of the images is almost imperceptible, and is undetectable in the case of individual objects. Nevertheless, by taking into account a very great number of galaxies, the effect can be detected, and reveals the matter content of the universe, especially the amount of dark

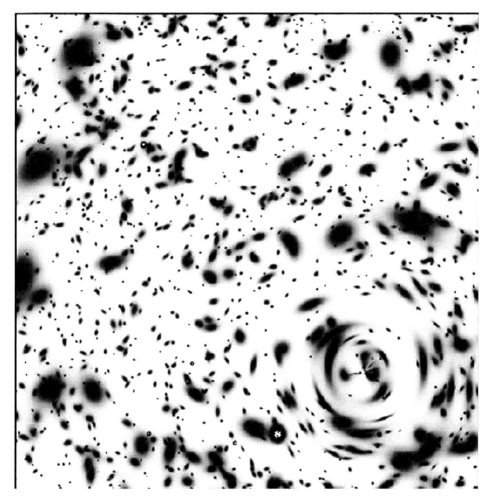

Figure 5.3 Gravitational shear. Simulation of deformations in images of background galaxies by gravitational lenses in the line of sight, and especially by a cluster of galaxies at bottom right. The deformations consist essentially of a tangential shearing of the images around the center of mass of the lenses. Using these deformations, we can measure, statistically, the ellipticity of thousands, even millions of galaxies, enabling us to establish a cartographical picture of dark matter projected onto the sky. An advantage of this method is that we can use it to determine the total mass of structures, independently of any possible bias due to galaxies (after Bernardeau and Mellier, 2003).

matter (Figure 5.3). This type of observation, ever more frequent nowadays, will also lead to the measurement of the parameters of dark energy.

Although clusters of galaxies contain only a very small proportion of the matter in the universe, they are representative objects, at the boundary between the domains of galaxies and large structures. The scale of clusters is that beyond which the visible/dark matter bias tends towards one, i.e. when the amount of visible matter is very similar to that of dark matter, and their densities are proportional. On larger scales, it is logical to assume that the effects of dissipation, star formation, and all the other associated complex processes no longer influence the dynamics of matter. It is precisely on these larger scales that the CDM model scores its greatest successes. Numerical simulations can also show that it is no longer possible on large scales to produce the segregation of dark and visible matter.

Globally within a cluster of galaxies, the ratio of baryons to non-baryons should have its universal value of one-sixth. The amount of mass observed in clusters confirms this prediction. For the most part, the visible mass exists in the form of very hot gas, emitting X-rays, and having ten times as much mass as that of the stars in these galaxies. Generally, in the universe as a whole, visible baryons (essentially within stars) account for only 10 percent of all baryons. In clusters, the rest of the ordinary matter, usually still in the form of gas too cool or too diluted to be visible, becomes visible in X-rays in the form of hot gas. By means of gravitational lenses, and through studies of the hydrodynamic equilibrium of hot gas, which gives information about the form and amplitude of the potential well that keeps the hot gas intact, it has been possible to measure the quantity of dark matter; it is indeed of the order of six times more abundant than visible matter. Even if we take into account that there remain considerable uncertainties, it seems that there cannot be any great quantity of dark baryons in clusters. No doubt the gas (warm or cold) containing them was heated by shock waves when the cluster was formed.

Another effect that confirms the CDM/dark energy model was detected in 2005, by correlating maps from WMAP and the Sloan Digital Sky Survey. This involves the blue-shifting of photons passing through large structures, and is called the Integrated Sachs-Wolfe (ISW) effect. Normally, in a flat universe dominated by matter (and not by dark energy, as our universe seems to be today), photons both enter and leave structures at the same level of energy. A large structure such as a supercluster of galaxies, about 500 million light years across, creates a certain gravitational potential well. Photons entering the well ought to gain kinetic energy, since they are losing potential energy. But, unlike massive particles, which accelerate, photons move at a constant speed. The energy gained resides therefore in the frequency of the photons; this increases, meaning that the wavelength contracts, causing the photons to appear bluer. If the supercluster of galaxies does not evolve during the time taken for the photon to traverse it, its exit is symmetrical and the photon gives up energy as it leaves, its frequency reverting to the initial value. However, in a universe dominated by dark energy, which causes an acceleration of the expansion of structures, the potential well will be less deep as the photon exits, and it will then be bluer (Figure 5.4).

The structure that gives rise to the potential well has its origins in a large density fluctuation, which must be visible in the anisotropies in the temperature of the cosmic microwave background. This idea led to the procedure of the

The problem of dark matter 129

Figure 5.4 Illustration of the Integrated Sachs-Wolfe (ISW) effect. Photons from the cosmic microwave background enter the potential well of a supercluster, on a very large scale (symbolized by the unbroken line). The time required for the photon to traverse the supercluster is of the order of 500 million years. In its descent into the well, the photon gains energy. If the potential well remained unchanged throughout the time required for this journey, the photon would lose exactly the same amount of energy as it left. However, because of the accelerated expansion of the universe due to dark energy, the galaxies of the supercluster are receding from each other and their potential well becomes less intense (dashed line). The energy budget is therefore positive in the case of the photon, which exits 'bluer' than it was on entering. This effect has been observed in the direction of super-structures, in the context of correlations between the WMAP map of anisotropies and the Sloan Digital Sky Survey of Galaxies.

detection of density fluctuations through the correlation of two maps. The signal of any one fluctuation is too weak to be individually detected, but the correlation of the two maps, covering 5,300 square degrees, has in fact detected and confirmed that the photons gain energy on leaving large-scale fluctuations. This detection has been carried out at various redshifts, and further results will come from the vast spectroscopic surveys planned for the future. Even today, these findings mean that we can eliminate certain dark-energy models.

Dark matter and the scaling relations of galaxies: the Tully-Fisher law for spirals

In 1977, two American astronomers, R. Brent Tully and J. Richard Fisher, discovered a correlation between the velocity width of the atomic hydrogen (HI) emission line (at a wavelength of 21 cm) and the absolute luminosity of a galaxy (Figure 5.5). This relationship had been proposed essentially as an indicator of distance, as the HI velocity width is independent of distance; measuring it using the 'Tully-Fisher relation' gives the intrinsic luminosity of a galaxy, and the distance of the galaxy can then be found from its apparent luminosity.

The Tully-Fisher relation is a power law linking the luminosity L of a spiral galaxy and its rotational velocity V_{rot}, such that $L \sim V_{rot}^4$. The form the relationship takes depends somewhat on the color in which the luminosity has

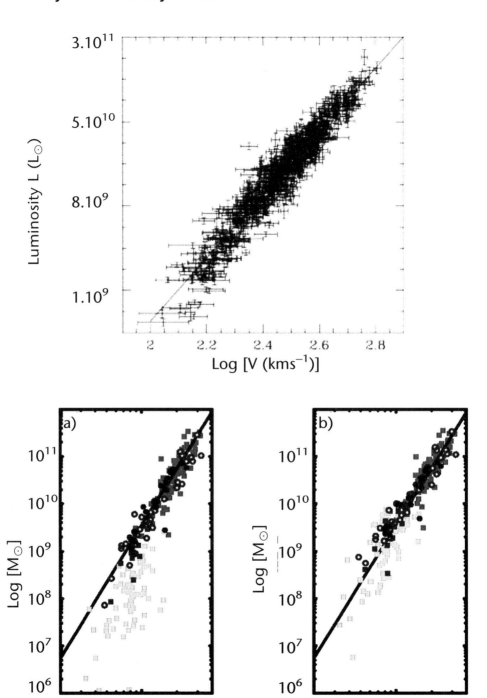

been measured. Originally, for large numbers of galaxies, observations could be made only in blue or visible light, using simpler techniques. In those colors, dispersion around the relation is quite wide. In fact, in blue light, the luminosity of galaxies depends largely on the recent rate of star formation, and on absorption by dust. The near-infrared band is however almost independent of dust, and far more stable in the presence of episodes of heightened star formation: in this band, radiation from old stellar populations dominates. As well as being useful as a distance indicator, the Tully-Fisher (TF) relation is a good indicator of the mass/luminosity (M/L) relationship in galaxies, and of its evolution.

So, what phenomena are at the origin of this relationship? Part of the relationship arises from the gravitational equilibrium of galaxies, and from the equality of potential energy and kinetic energy. The rotational velocity may be expressed as a function of the total mass M and the characteristic radius R, as $V_{rot}^2 \sim G M/R$, where G is the gravitational constant. This relationship must be completed by a correlation between mass and radius, derived from the history of galaxy formation. The relationship is also almost a power law linking the mass and size of a galaxy, in the form $M \sim R^2$. It may be interpreted as a quasi-constant average surface density for galaxies.

It was Ken Freeman who, in 1970, worked out that the surface brightness of the disks of spiral galaxies is, surprisingly, just about identical at their centers, for all galaxies. This constancy in stellar surface density is not merely an artifact caused by the absorption of light by dust, which would affect the brightness of the most massive galaxies, with their high metallicity and therefore abundant dust: the relationship is equally valid in the infrared, which is less prone to absorption. The universality of Freeman's law holds only in the case of fairly luminous galaxies. Now that it is possible to detect galaxies of very low surface brightness, Freeman's law is no longer verified for dwarf galaxies (Figure 4.16).

The two relationships described above, for giant spiral galaxies, point towards an explanation of the origin of the Tully-Fisher relation. What, however, is going

Figure 5.5 (opposite) Tully-Fisher relation for spiral galaxies. **Top:** Tully-Fisher law relating the rotational velocity (horizontal axis, log scale) to the luminosity of the galaxy in band I (vertical axis in units of solar luminosity). The rotational velocity is obtained from the width of the profile of the 21-cm HI line, corrected for the inclination of the galaxy (after Giovanelli et al., 1997). **Bottom left:** This has now been transformed into a velocity-mass relationship, by selecting a constant mass/luminosity for all the galaxies (see middle diagram). Light grey dots correspond to dwarf galaxies, with low rotational velocities, containing large amounts of gas. However, only the mass of the stars has been taken into account in this diagram. **Bottom right:** The mass of the gas is added to that of the stars, to give the total baryonic mass of the galaxies. The Tully-Fisher relation again appears. The straight line is a power law of slope equal to 4 (after McGaugh et al., 2000).

132 Mysteries of Galaxy Formation

on in the case of dwarf galaxies, with their lower surface brightness? In terms of total luminosity, they exhibit a deficit, clearly seen on the velocity-luminosity diagram (Figure 5.5). Nevertheless, it is possible to bring them back onto the common curve, by taking account of all components of visible matter. The reason is that these dwarf galaxies are very rich in gas, and their gas, unlike that of giant galaxies, constitutes a significant part of the total mass of the galaxy. By converting their luminosities into stellar masses, using the M/L relationship appropriate to the stellar populations according to the frequency band in question, the total mass, including the masses of both gas and stars, can be taken into account. Thus, dwarf galaxies satisfy the same Tully-Fisher relation as galaxies of high surface brightness. It would appear that this law applies to the collapse of baryons in dark-matter potential wells, and not to phenomena associated with star formation. This version of the relationship extended to all the visible mass of the galaxy is known as the baryonic Tully-Fisher relation.

Dark matter and the fundamental plane of elliptical galaxies

An analogous law applies to the elliptical galaxies. These exhibit little or no rotation, but their dynamic is represented by their velocity dispersion σ. The 'Faber-Jackson relation' relates total luminosity to the dispersion σ with a power law of exponent 4. Since the relationship is more dispersed than for spirals, it is necessary for elliptical galaxies to have recourse to their three principal parameters, luminosity L, dispersion σ and characteristic radius R. This relationship positions elliptical galaxies in a fundamental plane within their three-dimensional volume. This fundamental plane is not necessarily reduced to a straight line in projection on the plane L–σ; other combinations of axes must be sought, on the basis of variables that are merely combinations of the three fundamentals, with a view to reducing the dispersion. So it is possible to see the fundamental plane edge-on, in projection on two axes, one being the radius R and the other a combination of σ and the surface brightness L/R^2 (Figure 5.6).

Comparing these relations, we deduce that the mass/luminosity ratio varies with the luminosity, as $M/L \sim L^{1/4}$. The largest galaxies have a larger M/L, a result of their older stellar populations. This also corresponds to the separation in sequence between 'blue' and 'red' galaxies, where the age of the stars increases with the mass of the galaxies.

Elliptical galaxies are the natural result of mergers between two spiral galaxies, and more generally of successive mergers between several dwarf galaxies or spirals. Given that, today, they consist mostly of old stellar populations, it is probable that, at the beginning of the universe, these mergers created galaxies rich in gas, capable of forming stars at the time of interaction. Today, it is rather a case of galaxies merging at a late stage of their evolution, with little gas involved. Simulations have shown that the system resulting from this type of galaxy merger would still belong on the fundamental plane.

What is the fraction of dark matter around elliptical galaxies? This is a difficult

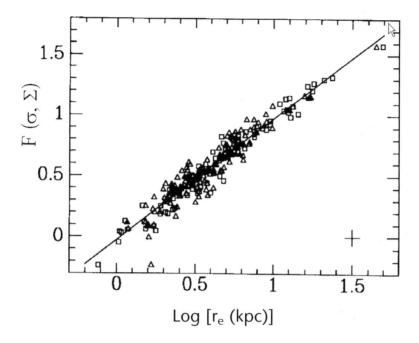

Figure 5.6 Scaling relations for elliptical galaxies. Fundamental plane of elliptical galaxies, seen practically edge-on in projection on this diagram relating the radius of the galaxy r_e, velocity dispersion σ, and surface brightness Σ. These parameters relate the visible mass (Σ) and the dark mass (σ) to the size of the objects (r_e). These parameters are confined to a plane, projecting as a straight line, in the adequate axes of projection shown. The position of the galaxies in this plane tells us a lot about the processes of the formation of elliptical galaxies, for example through mergers of smaller galaxies.

question to answer. With spiral galaxies, the geometry is simple: it is that of a thin disk, easy to deproject from the plane of the sky, and containing cool interstellar gas, which can be used as a tracer of the rotational curve. Elliptical galaxies have no atomic hydrogen gas, and gravitational potentials cannot be studied very far out from the center. Their geometry is ellipsoidal, either 'prolate,' i.e. the axis of symmetry being the major axis, or 'oblate,' i.e. the axis of symmetry being the minor axis. In most cases, there is in fact no axial symmetry, but the three axes are very different; these are the 'triaxial' galaxies.

These complex geometries result from their mode of formation, i.e. mergers of smaller galaxies, with randomly oriented symmetry axes. After several mergers, the final system gradually loses its angular momentum and its equilibrium no longer depends on its rotation, but on the velocity dispersion. Flattening in one direction is not due to rotation, as it is with the collapse of a gaseous system (giving rise to a disk), but rather to the fact that the galaxies that merged were in relative motion, in an orbit perpendicular to this direction. The velocity dispersion of the stars in the final system is therefore anisotropic, being greater in the direction of the relative orbit of the parent galaxies. It is the forces of 'pressure' resulting from this velocity dispersion that compensate for the gravitational forces, and ensure the equilibrium of elliptical galaxies.

One way to determine the distribution of dark matter is to measure the radial distribution of velocity dispersion, in order to make a dynamical model of the system, taking account of the stellar mass, with realistic values for the M/L relationship of the stars. This method reveals a very small quantity of dark matter, almost non-existent, within the optical radius of the galaxy, to the distance beyond which measurements of stellar velocities are no longer possible. There is much uncertainty here, as there are a great number of possible and equivalent solutions for deprojecting the galaxy, finding the values of the three axes, or the degree of anisotropy in the velocity dispersions as a function of radius. The effects of projection are complicated by the fact that the ellipticity also varies with the radius, as does the density.

During the last decade, the measurement of very weak velocity dispersions in the outermost areas of elliptical galaxies, thanks to the use of a new tracer, i.e. planetary nebulae, has sparked a debate about the presence or absence of dark matter in ellipticals. Planetary nebulae, such as the Helix Nebula in Figure 5.7, are envelopes of hot gas created in the expulsion of gas by old stars not much more massive than the Sun. Their kinematics are observed by virtue of the (relatively strong) ionized oxygen line [OIII] at 5007 Å, enabling us to observe the velocities of stars much further away than was possible before, using absorption lines. The observation that velocity dispersion falls off at a characteristic distance of three radii, although dark matter models predict a constant dispersion, came as a surprise. If spiral galaxies lie within large halos of dark matter, and become ellipticals by merging, how could the dark halos have disappeared during the mergers?

Now astronomers are not lacking in imagination, and immediately saw a solution to the mystery: the velocity dispersion becomes very anisotropic in the outer regions. Well away from the center, the velocities become almost radial, and even of very low amplitude, because the stars are nearly at the point of reversing the direction of their orbits. This phenomenon is amplified by the fact that the stars of the outer halo are formed, as the parent galaxies merge, principally during the first passage towards the pericenter. This causes the orbits of these stars to be particularly eccentric and almost radial, with pericenters very close to the center of the final elliptical galaxy, while their apocenters are very distant from the center. Numerical simulations show that it is therefore possible to bring down the velocity dispersion measured in projection very far from the effective radius of the galaxy. This is of course true only for stars, and not for particles of dark matter or gas, but these particles are either absent or invisible.

Measuring the velocities of stars is therefore too uncertain a method for determining the distribution of dark matter in ellipticals. We must resort to other methods, such as those involving the emission of X-rays from very hot gas in very massive galaxies, the hydrostatic equilibrium of the gas, or weak

The problem of dark matter 135

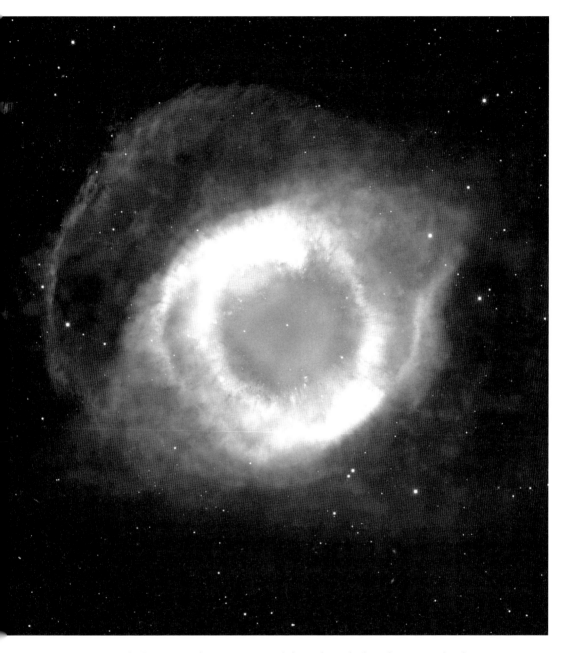

Figure 5.7 Hubble Space Telescope image of the Helix Nebula, a fine example of a planetary nebula. The faint speck at its center was once a star of greater mass than our own Sun. Now, near the end of its life, it has ejected its outer layers into space, and the remnant is a tiny white dwarf star, about the size of the Earth (The Hubble Heritage Team (STScI/AURA) and NASA).

gravitational lensing, statistically, using a large number of galaxies (though not restricting ourselves to one particular type); or, again, we could measure the velocities of companion galaxies in a group. Unfortunately, all these techniques are still in their infancy, and do not succeed in yielding accurate estimates of HI gas in spiral galaxies.

Has the ratio of dark matter to visible matter evolved over time?

The rotational velocity of spiral galaxies (or dispersion in elliptical galaxies) is directly linked to the amplitude of the potential well of the dark-matter halo in which the visible baryons are bathed. The correlations between luminosity (or mass) and rotational velocity observed for these galaxies have been established locally, at zero redshift. One way of testing the evolution of galaxies and the progressivity of the differential collapse of baryons and dark matter is to try to establish the equivalent of these relations as a function of redshift.

This is a difficult task, and only a few preliminary results have been obtained. Certainly, difficulties arise because of the lack of sensitivity of our current instruments when observing the very faint objects that remote galaxies have become, together with all the selection effects involved. On average, the surface brightness of galactic disks appears brighter than it is in reality (less bright disks are undetectable and therefore do not figure in the sample). Once this factor has been corrected for, it seems that evolutionary effects are not very perceptible, at least out to redshift $z = 1$. Similarly, galactic disk sizes seem to be smaller as we look back in time, but again this is due to the minimum surface brightness that the instrument can see; when the bias effects are corrected for, a fairly slow evolution is detected.

It is possible to show evidence of a more rapid evolution in certain categories of galaxies, among them the smaller 'blue' galaxies, while the massive 'red' galaxies do not evolve. The observed slow evolution of the Tully-Fisher relation suggests that galaxies between $z = 1$ and the present time evolve jointly by simultaneous accretion of gas and dark matter; thus, statistically, their position on the velocity-luminosity (or mass) curve stays the same. Elsewhere, it is indicated that there is a more marked evolution in clusters of galaxies: star-forming spiral galaxies in them tend to be brighter than in the field, for a given rotational velocity. On the other hand, lenticular (S0) galaxies in clusters appear less bright, and could be the final stage in the evolution of spirals that have lost their reservoir of gas. These preliminary results should be approached with caution, given the presence of very large selection effects.

The scaling relations verified for galaxies (the Tully-Fisher relation for spirals, and the fundamental plane for ellipticals), constitute an indispensable tool in our observations of the main physical properties of galaxies, and in particular the link between their dark matter and baryons. Statistically, the study of these relations has enabled us to highlight three major problems involved with the CDM model on the galactic scale.

The first major problem with the CDM model: cusps

Cosmological simulations are the principal tool in our investigations of the physics of dark matter, which is thought to consist of fairly massive non-baryonic particles that, on decoupling from photons in the early universe, were not relativistic (see Chapter 1). The initial conditions of the simulations are, for the very early universe, a homogeneous medium of particles, with primordial density fluctuations, corresponding to those observed in the cosmic microwave background by spacecraft such as COBE, WMAP and Planck. As a first approximation, it is possible to ignore baryonic matter, which constitutes only about five percent of the matter content of the universe. However, expansion is controlled by dark energy, which, to simplify matters, is considered to be a cosmological constant Λ, given that all observational constraints have so far been compatible with this hypothesis.

The values adopted agree with the latest observational results, i.e. a flat (= without curvature) universe where Λ = 73 percent, and Ω_m = 27 percent (see the Appendix on the various components of the universe). Of course, only a small part of the universe, a limited cube, can be simulated with computers, but the conditions at the edges of the cube are thought to be periodic, and the cube is repeated to infinity, to obviate edge effects. The size of the cube corresponds to the largest structures the simulation can handle. Also, the spatial resolution of the simulation corresponds to the smallest structures considered, and at present the ratio in size between the largest and smallest structures is approximately three orders of magnitude, not taking into account the possible zooms (cf. the biggest simulations ever undertaken, in a three-dimensional grid, with 2048 cells on a side, representing a volume of 2048^3 cells, or nearly 9 billion points). It is of course possible to reduce the size of the cube in order to achieve a better resolution in terms of the structures resolved within each halo, or galaxy.

The predictions of the CDM model for the typical radial profile of dark-matter halos have been determined in this way. Since these structures develop in much the same way no matter what scale is considered, this radial distribution is universal, for clusters of galaxies or for the galaxies themselves (if we ignore the baryons). This radial profile is a power law, as shown in Figure 5.8, no matter how far the spatial resolution is extended. This kind of shape leads to a singularity at the origin, and resembles what is known as a 'cusp': a density peaking sharply at the center. This cusp model conflicts with the 'core' model, which best fits with the observations. In the core model, density flattens out towards the center, below a certain characteristic radius, known as the core radius.

The predictions of the CDM model do not therefore correspond with observations, especially for dwarf galaxies, which are dominated by dark matter to a greater degree than other types. These galaxies show a flat density at their centers, with a very marked 'core' radius. The radial distribution is deduced from rotation curves, obtained using the tracer of atomic hydrogen gas. This is a component prone to dissipation, with a very low dispersion velocity, and it

138 **Mysteries of Galaxy Formation**

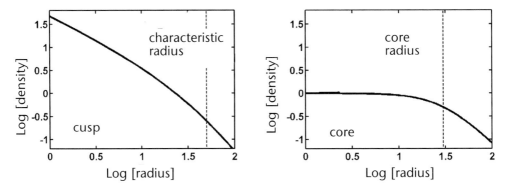

Figure 5.8 The cusp problem, predicted in simulations, when observing galactic cores. **Left**: radial profile of dark matter in cosmological simulations (CDM model). The density increases very rapidly towards the center, as the radius decreases. **Right**: radial profile of dark matter as observed in galaxies, based on rotational curves of the gas. The density no longer increases towards the center, but reaches a plateau. This central component is known as the 'core,' and is characteristic of an isothermal distribution of particles (i.e. constant velocity dispersion).

closely follows circular orbits in an axisymmetric potential. There are now so many nearby galaxies in which it has been possible to determine the velocity of the gas with sufficient accuracy and resolution, that a 'core' profile has been established and the predictions of the model have been contradicted.

One solution would be to consider the interaction of the dark matter with the baryons, which could run counter to the initial profile. Many hypotheses have been tested out using numerical simulations, in an attempt to flatten the density of the dark matter at the center, and dispose of the cusps, but to date, success has been limited. The idea is to provide the central dark matter with a good deal of dynamical energy, thereby reducing the depth of the potential well. The center could be heated through dynamical friction, by either a barred density wave or the formation of clumps of, for example, star clusters or interstellar clouds. Star formation could also provide extra energy to reduce the coherence of the central potential. However, in most cases, even this amount of energy seems insufficient.

Over and above the problem involving the radial profile of the CDM distribution, there is an equally serious one: that of the quantity of dark matter predicted by the models, which is far greater than that observed in giant spiral galaxies such as the Milky Way. This problem becomes particularly acute in the case of massive galaxies, where, as observations reveal, the fraction of dark matter is proportionally much smaller. Evidently, this conclusion is dependent on the stellar mass/luminosity ratio adopted, but there are now sufficient statistical data to confirm the existence of the problem: the M/L ratio of the stellar component can be estimated not only through the type of stellar populations observed, as a function of their color distributions, but also through the particular dynamical features of the disks.

To have the CDM models agreeing with the observations, the M/L ratio of the stars would need to tend towards zero, a case which is excluded. Neither can the ratio vary in too extreme a fashion from one galaxy type to another, without provoking too strong a dispersion in the observed Tully-Fisher relation. Inside galaxies of the Milky Way type, the models predict ten times more dark matter than has been deduced from observations, even if we ignore the fact that the falling of baryons into the potential well will lead to a contraction of the initial dark matter, thereby aggravating the problem further.

The second major problem with the CDM model: angular momentum

Another problem, doubtless related to the existence of the cusps discussed above, is that of the loss of angular momentum in galactic disks in cosmological simulations. For a given total mass, the disks of galaxies formed in simulations have a much lower rotational velocity than that is observed; in other words, the simulated galaxy disks are much smaller. Initially, however, the mean angular momentum is quite realistic, and corresponds to expected values, through tidal interaction and coupling of adjacent structures. The problem arises from the fact that the angular momentum of the baryonic matter disappears, to the advantage of dark matter, and very efficiently by means of dynamical friction, as galaxies merge. As galaxies interact, the angular momentum of their disks is transferred to the particles in the dark halos, and the baryons, merging and falling towards the center, are robbed of their rotation.

This problem is clearly seen in the Tully-Fisher relation, as described earlier. Compared with the observed relationship, that obtained in cosmological simulations certainly has the right slope, but is completely shifted, and the zero point is not reproduced, as Figure 5.9 shows. The slope is shifted downwards, i.e. for a given rotational velocity, it corresponds to a lesser mass; in other words, for a given mass, the galaxies in the model have too great a velocity, as they are too concentrated. The size of the disks is too small. The energy-related phenomena associated with the formation and evolution of stars (e.g. stellar winds, supernovae, etc.) have been used in an attempt to 'de-concentrate' the disks, but even when maximum values are employed, as in Figure 5.9, the problem persists.

The third major problem with the CDM model: satellite halos

Cosmological hierarchical-universe simulations have predicted the presence of a large number of sub-structures in dark-matter halos. The existence of a thousand sub-structures is in close agreement with observations in the case of the parent halo of a cluster of galaxies, so the same should be true for a halo as large as that of the Milky Way, for example. Our Galaxy ought to possess about 500 dwarf galaxies, similar in size to those already observed. The problem is that only a

140 **Mysteries of Galaxy Formation**

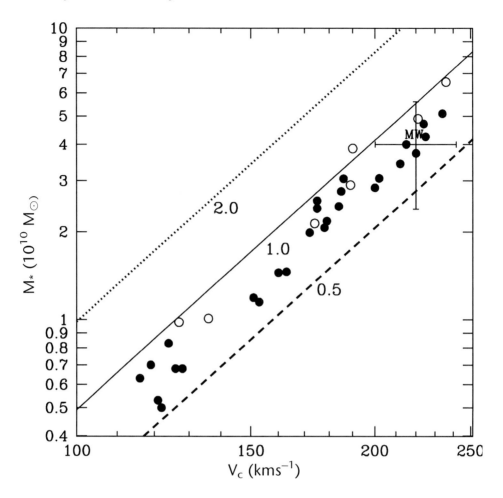

Figure 5.9 Tully-Fisher relation obtained in numerical simulations of dark matter (CDM) (filled or empty circles). Typically, the slope of the relation is well reproduced by the models, but the points are all below the line of the observed relation, which corresponds to the upper (dotted) line. Here, the three curves correspond to the relation of Giovanelli et al. (1997), where the M/L ratio in band I varies (between 0.5, 1 and 2, the value closest to reality). For a given mass, the rotational velocity obtained is too high, as galaxies have become too concentrated during collapse, and their radii are too small; but when star formation is taken into account, a greater concentration is avoided (after Sommer-Larsen et al., 2003).

dozen satellite galaxies have been found in the vicinity of the Milky Way, rather than the 500 predicted.

Figures 5.10 and 5.11 show the nature of the problem. Could it be that sub-halos are destroyed on a galactic scale, but not on the scale of a cluster of galaxies? We could surmise that dark halos exist around the Milky Way, for example, but are not filled with visible baryons, the stars having formed at

The problem of dark matter 141

Figure 5.10 Simulations of structures in the CDM model. The structure of a dark halo associated with a giant galaxy such as the Milky Way (left) or with a cluster of galaxies (right). In both cases, the edge of the image corresponds to the outer radius of the structures, where all the components are gravitationally bound (300 kpc in the left-hand image, and 2 Mpc in the right). The strong resemblance between these two structures shows that they are very similar across a great range of scales, a consequence of the independence of scale in the law of gravity. The results correspond well to observations in the case of clusters, but a problem arises in that of galaxies (after Moore *et al.*, 1999).

another time, and the action of supernovae having been violent enough to expel the vast majority of the baryons. Nevertheless, the presence of 500 sub-structures concentrated around our Galaxy ought to have been detected, from the dynamical effects of their destruction. Indeed, their passage through the disk should have heated it and eventually caused it to be destroyed. It might even be that the disk of the Milky Way could never have formed in the first place in the presence of such perturbations. More generally, if so many sub-structures really existed around all spiral galaxies, it would be difficult to explain the presence of thin disks of stars, certainly several billion years old.

Visibly, these sub-structures must have been destroyed at an earlier time. Again, the existence of cusps appears at the heart of the problem. If dark halos were less compact, they could have been more easily destroyed by tidal forces. Now, in cosmological simulations, the sub-structures are not ephemeral, but, on the contrary, remain very robust over billions of years.

142 Mysteries of Galaxy Formation

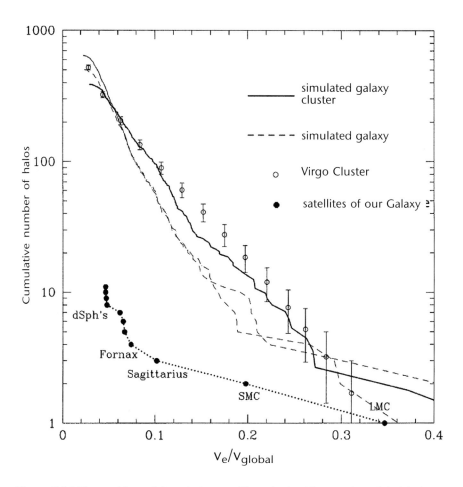

Figure 5.11 The problem of the missing satellite galaxies. The number of dark halos as a function of their mass M_h, or their circular velocity V_c ($V_c^2 = G M_h/R_h$, where R_h is the characteristic radius of the halo, containing the mass M_h), divided by the velocity of the parent structure (V_{global}). The curves originate from numerical simulations and are traced simultaneously for substructures within the halo of a typical giant galaxy such as the Milky Way (dashed curves, corresponding to two epochs 4 billion years apart) and for substructures corresponding to galaxies within a typical cluster of galaxies, such as the Virgo Cluster (unbroken curve). The filled circles represent observations of the dozen or so satellites of the Milky Way (some named), and the empty circles correspond to a mean for the galaxies of the Virgo Cluster. Note that the two dashed curves are equivalent, since evolution does not change the number of substructures (after Moore et al., 1999).

So what is dark matter?

Never, in astrophysics, have so many studies and articles been written on a subject about which more or less nothing is known for certain. There are those who claim that we are merely 'covering up' for our lack of knowledge about phenomena that are still barely understood. It is true that, as time has passed, the quantity of dark matter required has indeed diminished (along with our lack of knowledge?). For example, one of the first to resort to the idea of dark matter was the famous Swiss astronomer Fritz Zwicky, who, in 1933, measured the velocity dispersion of cluster galaxies, and showed that the visible mass was much too small to maintain the gravitational coherence of the cluster, given the observed velocities. Since then, the mass of hot gas, emitting X-rays, has been detected, adding ten times as much mass as that contained in the galaxies. What is more, it is true that today we can put quite precise limits on the quantity of dark matter required in the universe.

We even know perfectly well that there exist two sorts of dark matter: baryonic dark matter, since even today nearly 90 percent of all baryons have not been identified, and non-baryonic dark matter. The density of baryons in the universe has long been known, thanks to observations of the abundance of light elements, such as helium and deuterium, formed essentially at the time of the Big Bang. The total density of baryons in the universe today must be $\Omega_b = 4\text{-}5$ percent (see Appendix). Now visible matter contains only 10 percent of these baryons. Where might the rest be hiding?

Research has followed two main paths:

- The baryons may be confined within compact, non-radiating objects, such as the failed stars known as brown dwarfs (the remnants of stars at the end of their lives, such as white dwarfs, cannot contain much matter, because, during their lifetimes, they would have ejected heavy elements into interstellar space, and these are not observed in sufficient abundance). These compact objects have been sought around the Milky Way and the Andromeda Galaxy: astronomers have looked for the deviating effects they might have on light rays in their vicinity (gravitational microlensing), but too few of them have been found.
- The baryons may exist in the form of gas. This could be very hot, very diffuse gas in the cosmological filaments found throughout intergalactic space, or cold molecular gas. The hydrogen molecule H_2 is symmetrical, and does not radiate at the low temperatures of the interstellar medium; in fact, the ideal candidate for baryonic dark matter. This cold molecular gas may occur in the neighborhood of spiral galaxies, acting as a reservoir for star formation. It could easily explain rotation curves. But most of the cold gas could also lie within the cosmological filaments, 'sharing' the baryonic dark matter with the hot gas, as a multi-phase gaseous component.

The observation and quantification of the anisotropies in the cosmic microwave background, and the study of the signatures of gravitational lenses, supported by the standard candles which are the Type Ia supernovae, have enabled us to deduce the quantity of non-baryonic matter contained in the universe: 20 percent of the critical density required to close the universe, or to be more exact, $2 \cdot 10^{-30}$ g cm^{-3}. This is of course an average value, as the density is greater within the galaxies, i.e. of the order of 10^{-24} g cm^{-3} (the average for the Milky Way). Admittedly, the ultra-low figures reflect the fact that this matter has not yet been detected directly. But what are these WIMPS (Weakly Interacting Massive Particles), exhibiting practically zero interaction (unless it is gravitational) with the rest of matter?

It is probable that this unknown matter consists of various kinds of particles. First of all, we know today that neutrinos have mass. These particles interact only very weakly with other matter. Their mass has not been measured directly, but the phenomenon of neutrino oscillation, i.e. the oscillation between the three kinds ('flavors') of neutrino (electron neutrinos, muon neutrinos and tau neutrinos), proves the existence of this mass. Indeed, there can be oscillations only if the three flavors of neutrinos possess mass, and each has a different mass. The rate of change between the flavors enables us to deduce the differences in their masses, even though the mass of any one of them is not known. Current models fix an upper limit of 2.2 eV for the mass of neutrinos, so the neutrinos cannot contribute more than Ω_v = 12 percent to the contents of the universe. Nevertheless, this could represent up to 50 percent of non-baryonic matter! Great uncertainty remains, and the possible minimum value is only Ω_v = 0.4 percent.

The greater part of the mass must consist of 'cold' particles, non-relativistic since their thermal decoupling, forming large structures with spectra compatible with observations. The WIMP candidate that has sparked the most discussion is the neutralino, the lightest supersymmetrical particle. Supersymmetry (SUSY) is the theory where each particle of half-integer spin (or fermion) is related to one or more 'superpartners' of integer spin[1] (boson), while each boson is related to one or more 'superpartners' of half-integer spin. In this context, the neutralino is a particle left over from the Big Bang, and, although stable, could disintegrate as gamma-rays. Its mass is estimated at a minimum 40 GeV, but may be as much as a TeV (remember that the masses (m) of particles are expressed in equivalent units of energy (E= mc^2), the mass of the proton being 1 GeV. Supersymmetrical particles are far heavier than their partners, which explains why none of them has yet been created in a particle accelerator. Modern accelerators are not powerful enough to provide sufficient energy to create such massive particles: therefore, the theory of supersymmetry remains a hypothesis awaiting confirmation. The new Large Hadron Collider (LHC) at CERN (Figure 5.12) will

---

[1] 'Spin' is a quantum property associated with each particle. Elementary particles (fermions and bosons) are classified according to their spin. The spin that such particles carry is a truly intrinsic physical property, similar to their electric charge and mass.

operate at energies as high as 14 TeV, and may be able to throw some light on this question.

There is another much-studied hypothesis: the existence of a fourth kind of neutrino, the so-called 'sterile' neutrino. These interact with other neutrinos not via the weak interaction, but only through oscillations with them. They may not interact with other particles except through their mass and gravity, and are therefore dark matter candidates. The direction ('helicity') of their spin is right-handed (that of the other three kinds of neutrinos being left-handed); this is similar to the 'handedness' or chirality exhibited by certain molecules. The mass of these sterile neutrinos is totally unknown. They cannot have much mass, or they would at once disintegrate into several more common neutrinos, and, according to all constraints, the most likely mass is some tens of keV. Sterile neutrinos could be detected by their non-zero (if low) rate of disintegration, and the concomitant gamma-radiation.

Other dark-matter particles could be annihilated (with their associated anti-particles) and produce gamma-rays. This is in fact one of the preferred indirect methods for their detection. Some authors think that the high rate of detection of gamma-rays at 511 keV in the direction of the galactic center could be a manifestation of the annihilation of low-mass dark-matter particles. However, many astrophysical phenomena (e.g. supernovae) also produce this kind of radiation, as a product of electron-positron annihilation. Another (ultra-light) particle has been postulated in order to explain symmetry violation: this is the axion. The only possible mass permitted for the axion by the various constraints of physics and cosmology would be of the order of one micro-eV.

More exotic still are certain other theories that introduce extra dimensions to space. This is the case with the five-dimensional Kaluza-Klein theories, or with string theory. Originally, these theories were put forward in an (unsuccessful) effort to unify gravity and electromagnetism. The underlying idea is that gravity is the only force that can propagate in the extra dimensions, making it weaker than the others. Today these theories are still being developed with a view to resolving other problems, for example that of dark matter, which could be a visible manifestation in our space of the existence of these other dimensions. The universe might not be constituted only of the three dimensions of space and one of time, but may have other dimensions, variable in number according to the theory and symmetry they are thought to fit: just a few, or as many as dozens of additional dimensions. All the interactions with which we are familiar (the electromagnetic, the weak and the strong) are thought to confine themselves to the three dimensions; only gravity could extend into the others. Certain dimensions could be compact, i.e. folded in upon themselves, on microscopic scales.

In these current theories involving extra dimensions, the familiar four-dimensional space around us becomes a sub-space, a surface (or membrane), inside a higher-dimensional space. By extension, theorists have dubbed these sub-spaces 'branes.' The 'ekpyrotic' model, for example, holds that the Big Bang is the product of a collision between two branes. Our brane can vibrate in the

146 **Mysteries of Galaxy Formation**

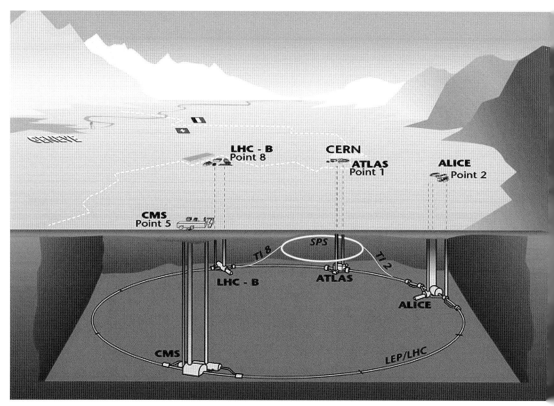

directions of the additional dimensions, and in quantum physics, these vibrations may manifest themselves in the creation of particles. These are stable particles, not interacting with others, and they are (of course?) known as 'branons': yet more of the many dark-matter candidates.

As we have seen, the theorists are never lost for ideas when they encounter problems in their observations of the universe! The great instruments of the next decade, be they accelerators or telescopes, will help them to untangle their hypotheses.

Figure 5.12 (opposite) The Large Hadron Collider (LHC) is the latest and most powerful in a series of particle accelerators that, over the last 70 years, have allowed us to penetrate deeper and deeper into the heart of matter and further and further back in time. The next steps in the journey will bring new knowledge about the beginning of our universe and how it works, as the LHC recreates, on a microscale, conditions that existed billionths of a second after the Big Bang. The LHC accelerates two beams of atomic particles in opposite directions around the LHC ring, which lies in a tunnel 27 km in circumference. When the particle beams reach their maximum speed the LHC allows them to 'collide' at four points on their circular journey. Detectors, placed around the collision points, are able to follow the millions of collisions and new particles produced every second and identify the distinctive behavior of interesting new particles from among the many thousands that are of little interest. **Top:** This diagram shows the locations of the four main experiments (ALICE, ATLAS, CMS and LHCb) that will take place at the LHC. Located between 50 m and 150 m underground, huge caverns have been excavated to house the giant detectors. The SPS (Super Proton Synchrotron), the final link in the pre-acceleration chain, and its connection tunnels to the LHC are also shown. **Bottom:** View of the LHC cryo-magnets inside the 27-km long LHC tunnel. It is vital that each magnet is placed exactly where it has been designed so that the path of the particle beam is precisely controlled (CERN, Geneva).

6 How can the problems be solved, and with what instruments?

How can we solve the current problems with the theory of galaxy formation?
The interaction between dark matter and visible matter involved in the violent phenomena of star formation; the energy released originating in the nuclear energy of stars; or even the energy phenomena related to the central black holes in every galaxy: can they prevent concentrations of dark matter?
And what if we had to modify the laws of gravity, in order to correctly represent observations on all scales?
What will the instruments of the future bring us?

Successes and problems: where are we now?

What is known as the 'concordance' model of the universe, with parameters drawn from a set of concordant proofs obtained through observation, has provided, during the past decade, a large number of remarkable results in our investigations of the formation of galaxies. For instance, we now know that:

- large structures were formed by gravitational collapse as a result of primordial fluctuations (of density and temperature) in the plasma issuing from the Big Bang), no doubt generated by the inflation during the first moments of the universe; their amplitude corresponds to what has been measured in the cosmic microwave background at the surface of last scattering;
- the existence of non-baryonic dark matter (CDM) is required to facilitate the development of the gravity potential wells, before the recombination era, and it also corresponds to the 'missing' mass. This is non-radiating, but is detected and quantified via the phenomenon of gravitational lensing, with light rays being deflected by all the matter in the universe;
- the standard ΛCDM model of the universe, based on cold non-baryonic dark matter (CDM) and dark energy (Λ), reproduces very satisfactorily, in numerical simulations, the power spectrum of the large structures, as well as the structure of the network of cosmic filaments, traced out by intergalactic gas and observed mainly by absorption in front of quasars.

However, when it comes to galaxies, the ΛCDM model predicts a quantity of dark matter far greater than that actually observed, and a too-concentrated density distribution in the form of a peak at the centers of galaxies, while it is density plateaux that are in fact observed. As a consequence, the baryons that fall into these potential wells must lose their angular momentum through dynamical friction, to the benefit of the halos, and the visible disks predicted by theory are far smaller than observations suggest. Finally, a large number of satellite halos surround each massive halo in the models, but these are much less numerous in reality.

In summary, the ΛCDM model has seen many successes on large scales (clusters and superclusters of galaxies), with their linear or semi-linear physics. However, on small scales (typically that of galaxies), where the physics becomes very non-linear, it encounters serious problems.

Dark-matter particles: self-interacting or colliding?

One of the solutions envisaged consists in supposing that particles of dark matter, although interacting only very weakly with the rest of matter, might display non-zero self-interaction. After all, we know nothing about these particles. According to the SIDM (Self-Interacting Dark Matter) hypothesis, the particles are deemed to have an effective collisional 'cross section', and even without the dissipation of energy, the scattering effect resulting from collisions could profoundly modify the radial distribution in dark-matter halos, and even prevent their concentration.

If the mean free path of the particles is short, of the order of the size of galaxies, the particles in the halos could be heated by shocks, and be unable to cool down again, and this would mean that concentration would be prevented. The collisional effect would be most visible on non-linear scales, and could eliminate the excess of small-scale structures. Numerical simulations have been widely used to test this hypothesis. Unfortunately, they have brought with them further difficulties.

One of the theoretical consequences of collisions between particles of dark matter is the spherical shape of dark halos (Figure 6.1). This is an aspect that seems not to correspond to the observations, which tend towards the conclusion that there is a significant flattening of the halos. In order to explain this flattening, it is necessary to suppose a greater initial angular momentum in the halos.

The degree of flattening in dark halos has recently been estimated thanks to various observational techniques. One of them takes advantage of weak gravitational lensing, and the deformations of background galaxies by a single foreground galaxy. Since these deformations are too slight to be detected individually, thousands of galaxies are 'stacked' and repositioned at the center of the 'mean' image, and the deformations are added together around this 'typical' galaxy. This method therefore measures the mean ellipticity of the halos, and tends to make them spherical if they are not oriented in the same way as the visible-matter galaxies situated at their centers, and which are used to align

How can the problems be solved, and with what instruments? 151

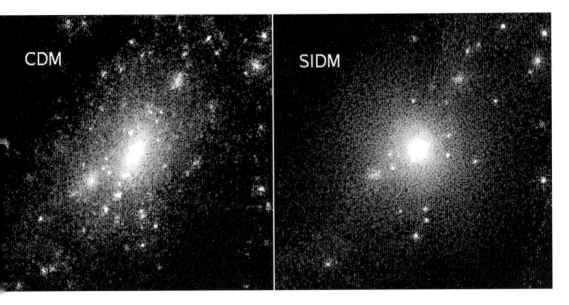

Figure 6.1 Numerical simulations of dark-matter halos. Results of classic CDM model without interaction (left), and of SIDM model (right) with particles able to collide with a non-zero effective cross section. Note that the model involving collisions is far more spherical, with fewer substructures, or small companion dark halos (after Moore et al., 2000).

them. The mean ellipticity found by this means is greater than 0.33, i.e. the ratio of the axes is less than 0.67. This ratio is in agreement with the CDM model, but not with the SIDM model.

Another method of measuring the flattening of halos comes from the study of polar-ring galaxies, in which the rotational velocity of the gas can be measured simultaneously in both the equatorial and polar planes. Yet another consists of examining the dynamics of disks of gas in the direction perpendicular to their planes: the gas opens out and warps, thereby tracing out the depth of the gravity well in this direction. These methods have not yet been applied to many galaxies, but the results so far tend to confirm the significant flattening in dark halos.

Collisions between particles of dark matter, if they existed, would be equivalent to a pressure exerted upon and within the halos. When a galaxy pursues an orbit, at a certain velocity, inside a larger halo, for example that of a cluster of galaxies, a dynamical-pressure phenomenon occurs; this is proportional to the square of the relative velocity. In a way, the galactic halo is 'peeled' by the wind exerted by the cluster of galaxies, and its radius is reduced to well within the tidal radius. This phenomenon is also seen in interactions between galaxies, and the SIDM models predict truncated halos with smaller radii than those of the CDM models. This prediction is a good argument for the SIDM model, because the observations tend in fact to show that dark halos of dwarf galaxies in the local group possess quite small radii, of about four times their optical radius, or half that predicted by the CDM models.

Besides, the extreme concentration of dark matter at the center of galaxies (the cusp problem, see Chapter 5), might be resolved by the introduction of particles of dark matter in collision, which would cause heating. However, if the radial distributions obtained in the simulations resemble, for a certain time, cores of constant density for certain values of the effective collision cross section, the problem is that the size of the core varies from one galaxy to another: in order to solve the problem of the distribution of dark matter in the dwarf galaxies, it would be necessary to select a value of the effective collision cross section adapted for each galaxy. Unfortunately, no single value of this effective cross section manages to reproduce all the observations.

Also, after a certain relaxation time, the core collapses and forms a cusp with an even more acute density law than before. Gravitational collapse becomes a runaway process. This stage represents a true gravothermal catastrophe, its more or less rapid progress depending on the value of the supposed mean free path of the dark-matter particles; and finally, the core collapses to form a black hole. In fact, the transition to a collisional physical mode serves only to exacerbate the cusp problem. Experiments with varying the effective cross section as a function of the velocity of the particles, or taking into account the continuous accretion of matter to prolong core heating, have shown that there can exist a small domain of parameter values enabling the construction of appropriate dark halos for a limited time, but equilibrium fails as soon as the baryons trigger the contraction of the dark-matter component.

Since changing the physical laws concerning dark matter has not resolved the problems encountered on the scale of galaxies, two paths towards a tentative solution suggest themselves. The first is the exploration of the complex phenomena involving baryons, phenomena that can modify the behavior of dark matter; the second, and the more far-reaching, is the exploration of modified laws of gravity. We shall examine each of these in turn.

A first avenue of approach: a better understanding of complex baryonic processes

One solution offering some hope, and much studied, is to suppose that energy phenomena affecting baryons could also secondarily affect dark matter, by heating the dense cores at the centers of galaxies, and reducing the density peaks by means of the ejection of some of the matter. These phenomena are negative feedback reactions, working against the concentration of matter. The latter process gives rise to starburst activity, which is self-limiting, as matter is then ejected: the young, massive stars react with stellar winds, due to radiation pressure on the dust, limiting their mass. At the end of their lives, massive stars explode as supernovae and re-inject a great deal of mechanical energy into the surrounding system.

How can the problems be solved, and with what instruments? 153

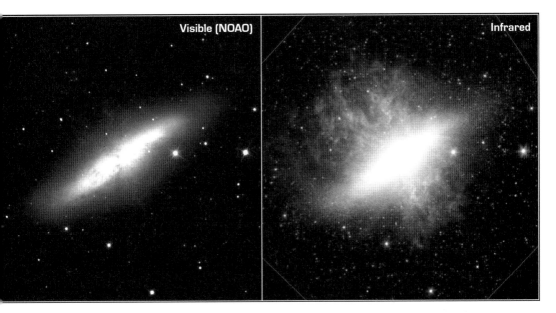

Figure 6.2 Starburst activity in the 'Cigar galaxy' M82, and a manifestation of self-regulating energetic phenomena. **Left:** visible-light image of the galaxy, seen edge-on, shows only a bar of light, mottled with dust lanes, against a dark patch of space. **Right:** mid-infrared image at wavelengths between 3.6 and 8 microns, from NASA's Spitzer Space Telescope. Old stars can be seen in the component seen edge-on, and dust is being ejected at right angles to the plane, in a bipolar flow, blown out into space by the galaxy's hot stars. Spitzer's infrared spectrograph told astronomers that the dust contains carbon-containing compounds called polycyclic aromatic hydrocarbons (PAHs) (Visible: NOAO; Infrared: NASA/JPL-Caltech/C. Engelbracht (University of Arizona)) See also PLATE 22 in the color section.

Dwarf galaxies are those most affected by these explosions, which eject gas at velocities of the order of 100 km/s. Galaxies of masses below about $3 \cdot 10^{10}$ M$_\odot$, with escape velocities of this order, can thus lose a large part of their interstellar medium. The gas from the star-forming region may be violently ejected from it, in a direction perpendicular to the disk of the galaxy, along the line of least resistance. Normally, the most violent episodes of starburst activity occur at the center of galaxies, and the gas is ejected in a bipolar flow, to either side of the galactic plane (Figure 6.2). This limits not only the formation of stars, but also the concentration of matter. This could be instrumental in flattening the radial distribution of dark matter.

Astronomers have seized upon this approach, running numerical simulations in an attempt to account for such energetic phenomena. Unfortunately, the energy involved in star formation is insufficient to resolve the problem completely. The ejection of a large fraction of the baryonic mass by superwinds of stellar origin indeed causes the dark matter to expand, but to an insufficient degree. It is also possible that turbulence engendered by star formation very early

154 Mysteries of Galaxy Formation

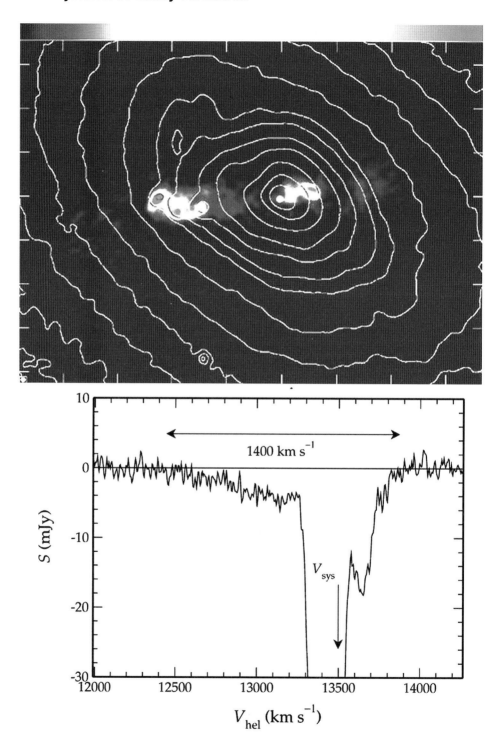

How can the problems be solved, and with what instruments?

on in the universe caused the interstellar medium to become clumpy, and, through dynamical friction, this might have flattened the profiles of dark matter, still being formed at this epoch. However, this cannot be generalized to include all dark-matter halos.

Since phenomena linked to star formation seem insufficient, and are above all limited by the total number of stars formed in the universe (to date, only a very minor part of the baryonic gas has been transformed into stars), astronomers are now turning to other high-energy phenomena such as active galactic nuclei (AGN), and the supermassive black holes that give rise to them, at the centers of galaxies. The regulatory phenomena exhibited by AGN are not limited by star formation, but they can intervene only within very massive, spheroidal galaxies, possessing massive black holes. This method thus complements that involving self-regulation due to star formation, itself more effective in dwarf galaxies.

A small number of AGN emit bipolar jets of ionized gas. These jets spread far out from the galaxy, like puffs of smoke; they are detected through their synchrotron emissions at radio wavelengths. These emissions of jets are one of the energetic manifestations displayed by quasars or radio galaxies, and occur intermittently. They can carry off a large fraction of the interstellar gas, and thus prevent star formation. They may also trigger the local expansion of dark matter. Figure 6.3 shows an example of radio jets emanating from the center of a spiral galaxy interacting with a companion, which, surprisingly, is not forming many stars, but expelling streams of gas. The plasma jets are not always oriented in the same direction, they can follow random paths as time passes, sweeping through a large area around the active nucleus. The heating of the surrounding gas offers an explanation of the moderation of cooling flows at the center of clusters (Figure 3.13). This represents a typical case of the self-regulation of the supply to

Figure 6.3 (opposite) Illustration of the influence of the active nucleus on the dynamics of the host galaxy. **Top:** emission of radio jets from the center of galaxy 3C293, in two lobes to either side of the center, with superimposed contours of the near-infrared image of the galaxy obtained by the Hubble Space Telescope (after Floyd et al., 2006). **Bottom:** absorption spectrum of atomic hydrogen (HI) at 21 cm, between the observer and the radio source 3C293, showing the existence of a flow of gas ejected by the galaxy at very high velocity, of the order of 1,400 km/s (after Morganti et al., 2003). The deepest absorption line at the center is due to the normal (rotating) gaseous component of the galaxy. The left wing of the spectral line corresponds to gas escaping at high speed towards the observer (absorption in the blue, blue-shifted by the Doppler effect). The galaxy harboring the radio source 3C293 is a spiral galaxy of peculiar morphology, which seems to stem from interaction/merger with a companion. Tidal perturbations are seen as dust lanes and in the perturbed kinematics of the gas. Surprisingly, the galaxy is not a site of starburst activity, despite an abundance of atomic and molecular gas. It could be that the activity in the nucleus, having given rise to the radio plasma jets, has sent the neutral gas out of the galaxy, in very rapid streams, preventing star formation.

supermassive black holes. Consequently, the processes regulating star formation and the growth of nuclei are intimately connected, providing an explanation for the relationship between the mass of black holes and that of spheroids.

Modified gravity

Now what if the supposed existence of dark matter were merely concealing some more significant modification of physical laws: a modification of the laws of gravity? This is certainly an avenue that will need to be explored quite seriously if no particle corresponding to non-baryonic dark matter is found by the new large particle accelerators.

Many propositions have been made involving modification of the form of the force of gravity, or that of the law of inertia. There is, however, only one that succeeds completely in explaining the rotational curves of galaxies, and, more generally, galactic-scale physics, in the manner of the Tully-Fisher relation. This is the MOND theory, proposed by Israeli physicist Moti Milgrom in 1983. MOND is an acronym for MOdified Newtonian Dynamics. Originally, this essentially involved an empirical modification of the form of the force, or gravitational potential, as a function of distance. The modification is suggested by the observation:

- that the rotational curves of galaxies (i.e. the velocity distribution according to radius) tend towards an almost constant value in the case of large radii, instead of decreasing in Keplerian fashion in the absence of mass (Figure 6.4);
- and that the radius at which this behavior is established varies from one galaxy to another, but always corresponds to a given value for acceleration a_0.

The crucial point here is that the common datum between the curves of all the galaxies, of very different types, is not a particular distance, but an acceleration. This is why the theories based on a modification of gravity as a function of scale, with characteristic radius, are doomed to failure.

Spiral galaxies, whose rotational curves are well known (thanks to the emission line of atomic HI gas at 21 cm), far out from the visible mass consisting essentially of stars, are of various kinds, and the fraction of dark matter required within them can vary quite widely. As a general rule, massive galaxies have proportionally less dark matter, and in effect, the critical value of acceleration a_0 is achieved at a greater radius, given their high mass. The small irregular dwarf galaxies contain a lot of gas, and their rotational curves show that they are dominated by dark matter. The efficiency of star formation in these small objects was so low in the past that, today, the mass of their stars is still far less than that of the mass of their gas. These dark-matter-dominated galaxies are an ideal laboratory for the study of dark matter, since uncertainty about the M/L ratio is here no obstacle to determining its influence. The mass profile is a direct tracer of the dark matter (or of modified gravity), without too much interference from other components.

How can the problems be solved, and with what instruments?

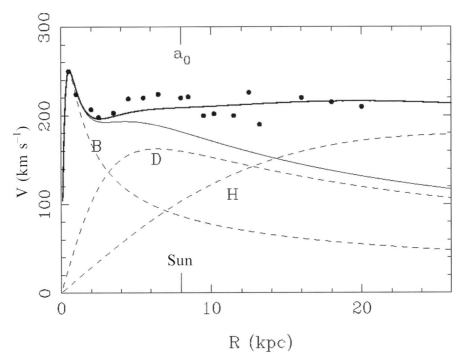

Figure 6.4 Rotational velocity of our Milky Way galaxy as a function of distance from the center (R in kpc). The Sun is situated at a distance of 8 kpc from the center. The filled circles indicate a compilation of measurements of neutral, ionized gas, or other stellar tracers. The measurements are very uncertain at great distances. The dashed curves indicate the modelled bulge contributions (B), and those of the disk (D) and the dark halo (H). The totality of these three components gives the bold curve, which adjusts the measurement points, with Newtonian gravity. The unbroken curve is the contribution of visible matter only (bulge plus disk). Note that for each radius, the sum of the contributions is found using the square of the velocities, in the Newtonian regime. The MOND critical acceleration a_0 is achieved at the distance of the Sun from the center. Thus, our region lies within an intermediate regime between Newtonian and MOND. The application of the modified law of gravity enables us to adjust the measurement points to the visible components only, without recourse to the addition of the dark halo.

The principle of MOND is that:

- when acceleration exceeds the critical value a_0, gravity is Newtonian, i.e. $g_N = GM/r^2$ (where G is the gravitational constant, and M the mass);
- beyond this value, the force does not decrease as $1/r^2$, but as $1/r$ (the potential is no longer $1/r$, but logarithmic). Acceleration in the MOND regime is $g_M = (a_0 g_N)^{1/2}$. This acceleration being always equal to V^2/r, it is easy to see that, in this relation, the radius disappears. Asymptotically, the rotational curves tend towards a constant V_{rot}, such that $V_{rot}^4 = a_0 GM$; which gives the Tully–Fisher

relation, if the mass-to-luminosity ratio (M/L) of the stars can be considered almost constant.

Of course, between the Newtonian regime, generally verified for the center of spiral galaxies, and the MOND regime for the outer parts, the system shows a continuous progression via an intermediate regime, and the schematic behaviors mentioned above are only asymptotic behaviors. Several versions of the force of gravity exist in this intermediate regime, and the exact form would have to be determined experimentally. The value of a_0 which reproduces the rotational curves is $a_0 = 1.2 \cdot 10^{-10}$ m/s^2, representing an extremely low acceleration, when we consider that, on Earth, the acceleration due to gravity is 10 m/s^2.

One of the great successes of MOND is that one single parameter, a_0, enables us to represent all of the rotational curves (Figure 6.5). By comparison, the ΛCDM model requires many free parameters to achieve the same result. In effect, for each galaxy, a variable quantity of dark matter has to be added and the radial distribution adjusted to the profile of the rotational curve. Since each galaxy has its own parameters, an infinity of parameters is needed. Many of the dynamical processes of galaxies have yet to be considered in the context of modified gravity. To date, the model reproduces galactic scales very well, more so than the ΛCDM model: it solves not only the cusp problem, and that of the amplitude of dark matter, and of how visible and invisible matter conspire to produce flat rotational curves, but also that of the stability and signature of spiral arms in the rotational curve, etc.

On the other hand, in the MOND regime, the dynamics of galaxies become less intuitive, since the law of gravity is no longer linear: the attractive force due to a mass A and a mass B together at one point is not the sum of the forces due to A and B separately. A Lagrangian formalism was very soon put in place to represent the MOND theory and satisfy all the laws of conservation. In this context, the motion of a composite object such as a star or a cluster of stars is independent of its internal acceleration, and is defined as the motion of the center of gravity in the external field, which can be accommodated in the MOND regime.

A more satisfactory justification of this model was obtained by Jacob Bekenstein in 2004. Bekenstein introduced a relativistic, covariant theory, involving a tensor-vector-scalar (TeVeS) field. This theory replaces Einstein's general relativity in the domains where the dark-matter problem dominates. For example, MOND is now capable of explaining the phenomena of gravitational lensing.

The MOND problem in clusters of galaxies

In spite of its remarkable successes on the scale of individual galaxies, there are still problems with MOND on the scale of clusters of galaxies. In galaxies, the effects of dark matter are encountered mostly in the outer regions. But in clusters

How can the problems be solved, and with what instruments? 159

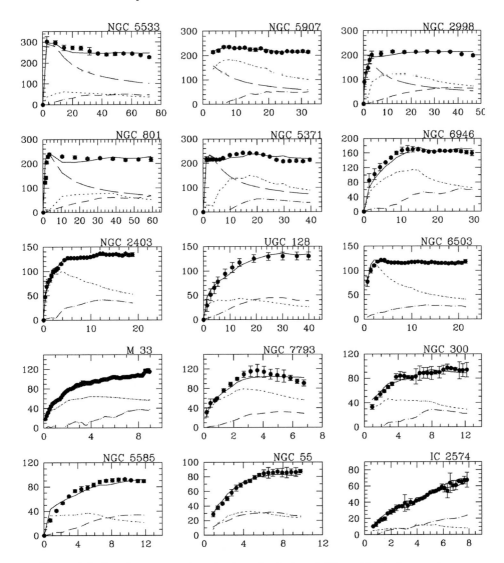

Figure 6.5 Rotation curves of a selection of galaxies of different types (name above each panel). The horizontal axis is the distance from the center in kpc (1 kpc = 3,260 light years). The vertical axis is the rotational velocity in km/s. The points represent velocity measurements (21-cm atomic hydrogen line) using the Doppler effect. Error bars are shown. The dotted and dot-dashed lines are expected (Newtonian) rotational velocities of the visible component of stars and the HI gas itself. The occasional dashed-only line represents the contribution of the bulge, if present. The solid lines represent the MOND predictions, taking into account the gas and stars, with a constant M/L ratio. Note that the galaxies in the upper panels are massive (high rotational velocity, greater than 300 km/s) and dwarfs in the lower ones (velocities not exceeding 80 km/s), and that the contribution of the gas and 'dark matter' increases from the upper to the lower panels. MOND manages to explain all the observations, with just one parameter (a_0), whatever the fraction of invisible matter, which varies greatly from one galaxy to another (after Sanders and McGaugh, 2002).

160 Mysteries of Galaxy Formation

Figure 6.6 Composite image of a collision between two clusters of galaxies. Superposition of X-ray emissions from hot gas and the projected mass in the galaxy cluster 1E 0657-56, familiarly known as the 'Bullet Cluster' (X-ray: NASA/CXC/CfA/ M.Markevitch *et al.*; Lensing Map: NASA/STScI, ESO WFI, Magellan/U.Arizona/D.Clowe *et al.*; Optical: NASA/STScI, Magellan/U.Arizona/D.Clowe *et al.*). The mass distribution projected on the sky corresponds to the mass of the clusters as reconstructed by gravitational lenses (deformations of background galaxies). Two clusters can easily be distinguished. The smaller cluster, at right, seems to have traversed, in the manner of a cannon ball, the larger one at left. During this collision, the hot gas of the subcluster has encountered the hot gas of the large cluster, and has been braked such that the two gaseous areas are closer together than the two masses. To the right we clearly see a shock wave in the shape of an arc, resulting from the passage of the smaller galaxy. The projected X-ray gas and masses are also superimposed on the optical image showing the individual galaxies. The different behavior during the collision of the hot gas and the stellar masses, including dark matter, allows us to separate the three components and test the models (after Clowe *et al.*, 2006). See also PLATE 23 in the color section.

of galaxies, the missing mass is detected mostly at the center, where acceleration is relatively high (i.e. in the Newtonian or intermediate regimes). Rich clusters of galaxies are very special regions of the universe, where nearly all the baryonic matter is visible. The gas constituting the dark baryons in the rest of the universe has been heated by shocks during the formation of the cluster, and represents more than ten times the visible mass of the galaxies. The dark matter in the cluster is delineated by gravitational lenses and by the hydrostatic equilibrium of hot gas, which emits X-rays.

These two methods come together to produce a concentrated profile, and the total mass of dark matter is five times that of the baryonic matter, as in the rest of the universe. The MOND model manages to reduce the requirement for non-baryonic dark matter, though not completely. The solution therefore consists in the involvement of neutrinos, whose exact mass is still unknown. By choosing a value compatible with current uncertainty (about 10 percent of the universe's neutrino content), we can explain the dynamics of clusters. Clusters of galaxies are not always in equilibrium, and sub-structures and sub-clusters merge to form larger objects.

A very atypical case of a collision between two clusters of galaxies has given rise recently to an animated debate on the nature of dark matter. In this structure (1E 0657-56, popularly known as the 'Bullet Cluster'), a sub-cluster is moving through a more massive cluster, creating a shock wave in the shape of a characteristic arc within the X-ray gas (Figure 6.6). The rapid passage of the sub-cluster though the cluster, at a velocity of 4,500 km/s, is causing hot gas to be displaced within the total mass, which can be mapped thanks to the effects of gravitational lensing on background galaxies. The hot gas is displaced almost outside its parent gravity potential well, contrary to the equilibrium situation where it coincides. Of course, the gas is not in equilibrium and it is difficult to know its mass, but it seems not to represent a dominant fraction of the mass; if it did, the cartography of the total mass would not be completely dissociated from it. This rare displacement between the various components gives a chance to appreciate the relative contributions. This example can be interpreted as proving the existence of non-baryonic dark matter, which does not follow the hot gas.

However, the Bullet Cluster does not seem to produce more problems with the MOND model than do other clusters. The MOND explanation of cluster dynamics already takes account of the existence of approximately 10 percent neutrinos, or twice the percentage of baryons. The neutrinos do not interact with the hot gas, and, since they traverse the ensemble of the galaxies without collisions, they separate from the gas. The observed deformation in the maps of mass in the direction of the gas is completely compatible with the MOND model.

MOND and the formation of galaxies

Unlike in the Newtonian regime, where non-baryonic dark matter is indispensable for the formation of structures and galaxies, the MOND regime's primordial fluctuations increase at a much faster rate, because gravitational attraction is stronger at great distance. It seems therefore that there has been enough time since the recombination of hydrogen (380,000 years after the Big Bang) to form the non-linear structures of today. The formation of the first structures and the first galaxies is still a largely unexplored field in the context of modified-gravity models. Preliminary calculations show that structures form very early on, and the first stars can re-ionize the universe. This corresponds to observations of the cosmic background and of absorption lines from distant quasars.

Might we be able to observe other consequences of modified gravity on smaller scales, closer to us? It could be that acceleration at the edge of the Solar System is of the order of the critical acceleration a_0. An abnormal acceleration was detected by the Pioneer space probes in the 1980s after their encounters with Jupiter. To be more precise, beyond a distance of 20 astronomical units from the Sun, their trajectories can be explained only by introducing an additional force of attraction of the order of 6 a_0 (or $8 \cdot 10^{-10}$ m/s^2). The amplitude is a little greater than is necessary to explain the rotational curves of galaxies; however, many uncertainties remain. The observation is based upon data acquired more than twenty years ago, and many hypotheses have been tested out since then, including an examination of the characteristics of the Pioneer probes themselves, e.g. gas leaks, etc. The mystery persists.

The modification of matter, not gravity

Another interesting possibility has been put forward by the group led by Luc Blanchet at the Institut d'Astrophysique in Paris. Their idea is based on the fact that the MOND equations, verified on the scale of galaxies, are absolutely analogous to the physical mechanism of polarization by an external field, as found in electromagnetism. The physics of the phenomenon is well illustrated by comparison to the immersion of a dielectric into an electrical field. A dielectric is a non-conducting substance (i.e. an insulator), through which current cannot travel, because all its electrons are attached to atoms. When the dielectric substance is immersed in an electric field **E**, the atoms are polarized, i.e. the nuclei of the positively charged atoms move in the direction of the field **E**, while the center of gravity of the negative charges, forming the electron cloud, moves in the opposite direction. This charge displacement is equivalent to a dipole p, aligned with the field **E**. The density of the dipoles in the material is equivalent to a polarization field, which in a manner of speaking is the opposite of the electric field. The effective field in a dielectric is weaker than it would be in a vacuum; this is what is known as the external-charge screening effect, creating the field **E** through polarization charges.

So the idea is to interpret the MOND effects as the modification of the equation of gravity by a 'digravitational' medium. We no longer postulate a modification of the law of gravity, but rather the existence of a new kind of dark matter, formed from dipole moments, aligned in the gravitational field. The screening effect here would be of opposite sign to our electric analogy, the 'screened' field being greater than the original field, unlike the electric field. This new matter is characterized by gravitational dipoles, which will normally not co-exist with ordinary matter. In effect, the electric analogy supposes that we are constructing a dipole with two 'charges' of opposite signs, or, here, a positive mass and a negative mass. But the analogy cannot be taken to its conclusion, as the dipoles are explained by the introduction of a fifth force, and the dipolar medium resembles a kind of static ether, as with the dielectric, whose charges are fixed to the atomic sites. This dipolar fluid obeys the equations of Einstein's general relativity, and has the advantage of including dark energy in the form of a cosmological constant.

... and how does string theory come into this?

Of course, whether or not dark matter and dark energy exist depends, as we have shown above, on the cosmological model used. Several approaches have shown that non-equilibrium models, in a time-dependent context, could reproduce the observations, while the conventional model is based on equilibrium. String theory, currently the best quantum gravity theory, also offers a solution in its non-equilibrium version. In fact, it is possible to find a formulation of string theory that reverts to the specifications of modified gravity. The common feature of these theories is the introduction of a vector field, violating the invariance of the laws as a function of the reference point, by imposing a privileged frame of reference. These theories are also accompanied by a scalar field, which is in a way a reintroduction of the isotropic 'ether' component of space. The ether is at rest in the privileged frame of reference. In certain contexts, string theory can lend some theoretical justification to the model, and relies on a geometry involving several dimensions (at least eight), the extra dimensions possibly being compactified, in order to represent our familiar four-dimensional space. In these models ordinary and massive neutrinos also play an essential part. Lastly, it should be noted that other models are based on the possible existence of massive 'sterile' neutrinos, or 'right-handed' neutrinos (of positive spin in the direction of their motion); the more familiar, ordinary neutrinos that we detect in accelerators and in space are 'left-handed.'

Elsewhere, in an attempt to deal with the problem of the re-acceleration of expansion (and therefore of the existence of dark energy), a certain number of 'degravitation' theories have been put forward, that reduce or suppress gravity on a very grand scale. One example of such a theory considers that gravitational interaction is mediated by massive gravitons, whose mass introduces a characteristic scale. Beyond this scale, the force of gravity no longer decreases as the square of the distance, but as the cube of the distance. These theories also include extra spatial dimensions, the idea being that gravity extends to the other dimensions, and is thereby weakened in the principal space: our own. Gravity is

164 Mysteries of Galaxy Formation

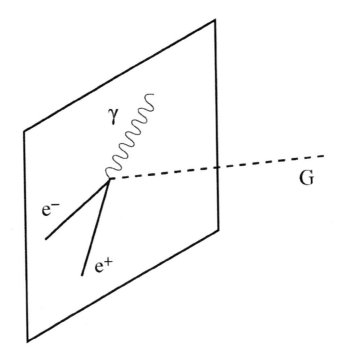

Figure 6.7 Schematic representation of a 'brane' model of the universe. The plane represents very succinctly the 4-dimensional universe (with its three spatial dimensions and one temporal) with which we are familiar. All interactions, especially electromagnetic ones, are confined within this space. Only gravitons can escape from it into other dimensions (after Cavaglia, 2002).

the only force able to penetrate the extra dimensions, while the other forces, such as electromagnetism, remain confined to our space (Figure 6.7). Each of these theories offers predictions capable of being tested by more detailed observations of the universe and its great structures.

Instruments of the future: ALMA, JWST, ELT, SKA …

Clearly, the enormous advances that have been made in recent years in our understanding of the beginning of our universe and the formation and evolution of galaxies have also led us into uncharted territory. We have realized that there is much that remains to be discovered: our surveys of the contents of the universe reveal to us that visible matter represents less than one percent of the total! One of the main priorities is that of unraveling the mysteries of the unknown components, dark energy and dark matter, and finding out if they entail drastic modifications of our physical laws. In order to properly determine the equation of state of dark energy, a whole battery of techniques will be

How can the problems be solved, and with what instruments?

deployed during the decade to come, both in Europe and in the United States. At least four major projects will be brought to bear on the problem:

- cartographic surveys of weak gravitational lenses, in tandem with the use of large-scale spectroscopic data;
- the study of baryonic acoustic oscillations, at various redshifts around $z = 1.5$-2, i.e. the epoch when dark energy began to dominate the contents of the universe;
- observation of a large number of Type Ia supernovae at the same redshifts;
- large-volume surveys of clusters of galaxies.

All these methods will require enormous statistical support in order to increase the accuracy of the values of the parameters, down to a level of one percent. Today, the equation of state linking the pressure and energy of dark energy is known to perhaps 10 percent, but many models are compatible with this range of uncertainty, for example the cosmological constant, the existence of a fifth element (quintessence), and many other possibilities. Whatever the tracer to be used, whether cartographic surveys of gravitational lenses, baryonic oscillations, clusters of galaxies or the 'standard candles' of supernovae, a large area of the sky will have to be observed, if only to obviate the effects of cosmological variance. In other words, we must make sure that we are not concentrating on just one 'special' region of the universe.

To detect and follow the light curves of supernovae, NASA is planning to launch a dedicated space telescope, called the SuperNova/Acceleration Probe (SNAP) of comparable diameter (2 m) to the Hubble Space Telescope. The new instrument will, however, have a wider field of view (0.7 square degrees), in order to cover a large area of the sky. The SNAP telescope (Figure 6.8) could also perform gravitational-lens cartography, and photometrically, by means of multiple filters, obtain millions of redshift measurements. It goes without saying that the instruments needed to secure such a vast volume of data will be gigantic, and the cost will be great.

Less expensive ground-based instruments are also envisaged, even though their imaging capabilities will be less impressive. Wide-field terrestrial spectroscopic telescopes of up to 10 m in diameter have also been proposed, but also the revolutionary SKA radio telescope (of which more later). The SKA will also be able to perform gravitational-lens cartography at radio wavelengths, across almost the whole sky. It will carry out precise hydrogen-line spectroscopy involving a billion galaxies, allowing us to quantify baryonic oscillations to an unprecedented degree.

As we saw in Chapter 2, there will be great advances in the detection and observation, at millimeter wavelengths, of infant galaxies in their cocoons at the beginning of the universe: an international collaboration involving the United States, Canada, Europe, Japan and Chile has been set up with a view to constructing ALMA (Atacama Large Millimeter/submillimeter Array), a network of more than fifty antennae on the Atacama plateau in Chile (Figure 6.9). The site will accommodate a 14-km array, allowing interferometry to be done using

166 **Mysteries of Galaxy Formation**

Figure 6.8 Artist's impression of the SuperNova Acceleration Probe (SNAP), which will be able to detect and observe several thousand supernovae out to redshifts of approximately $z \sim 1.7$. It will also be able to measure gravitational lensing effects created by the distribution of dark matter in the universe. (SNAP/LBL.)

the widest possible separation, with a consequent gain in spatial resolution. The frequencies used will range from 30 to 950 GHz, corresponding to wavelengths between 0.3 and 10 millimeters. The plateau is at an altitude of 5,000-5,500 m, making it second only to the South Pole as the best site in the world for astronomy. It is a remarkably dry area, and this is a vital requirement since, at the frequencies involved, the lines of water vapor in the Earth's atmosphere will set a

How can the problems be solved, and with what instruments?

limit on transparency. Most of the antennae are 12 m in diameter, and an auxiliary network of 7-m antennae will enable studies at certain spatial frequencies with a wider field of view. Correlation of the signals from all these antennae will engender an unprecedented number of operations per second: $1.6 \cdot 10^{16}$, or 16 petaflops.

Although these operations will take place on the high Llano de Chajnantor plateau, the main work base will be located at a lower altitude near the village of San Pedro de Atacama. ALMA will be the biggest telescope on Earth during the coming decade, and the world's largest network of radio antennas. One of its flagship programs will be a study of high-redshift galaxies, and of their formation during the earliest stage of the universe. Also, ALMA will be able to tell us much about the formation of stars and their protoplanetary environs, at the embryonic phase when objects were still hidden within their dusty cocoons. It will be able to detect the faintest objects, at all redshifts, by observing the thermal emissions from dust, and also give us information on the physics of objects, their motions and masses, gas content and the efficiency of star formation through the study of strongly redshifted molecular lines.

At the same time, distant galaxies will be observed at visible and infrared wavelengths, to study their content of stars and ionized gas. The very faintness of these objects will require the use of large telescopes, and a 42-m telescope is planned by European astronomers (Figure 6.10). At the same time, their American counterparts are investigating the possibility of a 30-m telescope (Figure 6.11). This new generation of telescopes can be realized only through the use of adaptive optics, correcting in real time for distortions of images due to atmospheric turbulence. At present, this kind of correction is done by comparing the wavefront of the light of the object studied with that of a bright star in the same field of view. However, this limits the field to only very small areas of sky, near reasonably bright stars. The correction can be extended to a large part of the sky only by using artificial 'laser guide stars', now in ever more general use by large telescopes.

Figure 6.9 (previous page) **Top and bottom:** Artist's impressions of the Atacama Large Millimeter/submillimeter Array (ALMA), one of the largest ground-based astronomy projects of the next decade, which will be a major new facility for world astronomy. ALMA will be comprised of a giant array of 12-m submillimeter quality antennas, with baselines of several kilometers. An additional, compact array of 7-m and 12-m antennas is also foreseen. Construction of ALMA started in 2003 and will be completed in 2012. The ALMA project is an international collaboration between Europe, Japan and North America in cooperation with the Republic of Chile (NRAO/AUI and ESO).

How can the problems be solved, and with what instruments? 169

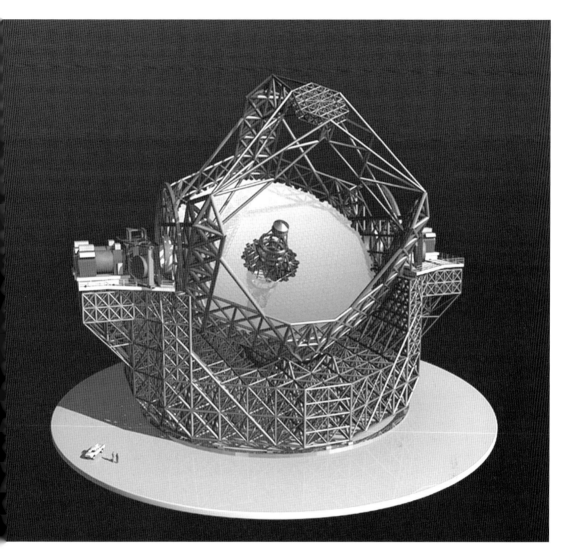

Figure 6.10 Artist's impression of the E-ELT (European Extremely Large Telescope), which is planned to have a primary mirror 42 meters in diameter, composed of 902 segments, each 1.45 meters wide, and a secondary mirror as large as 6 meters in diameter. The total rotating mass will be 5,500 tons. Its eventual location is still undecided. Its enormous size will enable it to probe the earliest galaxies and stars in the universe. Observations will be carried out using adaptive optics, meaning that wavefront deformations due to atmospheric turbulence will be corrected in real time by comparing them with light waves generated by a 'laser guide star'. (ESO).

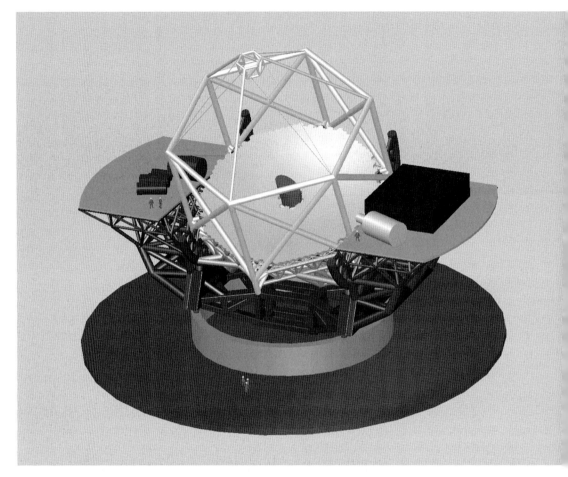

Figure 6.11 The Thirty Meter Telescope (TMT) is shown in this computer-generated illustration. The primary mirror consists of 492 separate hexagonal segments, each controlled by individual actuators. The design expands on the successful operation of the 10-meter Keck telescopes. A centrally-located tertiary mirror directs light to the Nasmyth focus, located along the horizontal axis of the mount (TMT).

The spectroscopic instruments attached to these giant telescopes are massive three-dimensional affairs enabling the creation of spectra for each point of the image, by steering the photons received by each pixel towards a spectrograph, using fiber optics or micro-mirrors. Just as the Hubble Space Telescope now collaborates with 10-m telescopes on the ground, producing respectively both images and spectroscopic analyses, the giant ground-based telescopes of the future will work with the planned JWST (James Webb Space Telescope). The JWST (Figure 6.12), a 6.5-m instrument, will work preferentially in the infrared. It will be launched by NASA in about 2013, and will be situated at the L2 Lagrangian point 1.5 million km from Earth.

How can the problems be solved, and with what instruments?

Figure 6.12 An artist's impression of the James Webb Space Telescope (JWST), a large infrared telescope with a 6.5-meter primary mirror. Launch is planned for 2013. It will study every phase in the history of our Universe, from the first luminous glows after the Big Bang, to the formation of solar systems capable of supporting life (NASA/ESA).

The SKA (Square Kilometer Array) is a project involving a radio telescope covering one square kilometer (one million square meters) working in the centimeter-meter domains. In total, there will be about a thousand antennae, in a collecting area of diameter about 3,000 km, though half of the antennae will be in a condensed inner array within an area 5 km across (Figure 6.13).

This enormous radio telescope will be 100 times more sensitive than our current instruments. It will operate at frequencies from 0.15 to 25 GHz (i.e. wavelengths from 1.2 cm to 2 m), and its field of view will be at least several square degrees at the wavelength of the 21-cm atomic hydrogen line.

172 **Mysteries of Galaxy Formation**

Figure 6.13 Artist's impression of the central area of the Square Kilometer Array (SKA). Two kinds of antennae will be used to cover a wide range of frequencies, at wavelengths between 1.2 cm and 2 m. The small, steerable paraboloids will work at high frequencies, and the fixed tiles at the center at low frequencies. All the antennae will be equipped with phased networks as receivers, enabling the electronic reconstruction of simultaneous observations in several directions (Xilostudios).

Thanks to new technologies based on the power and compactness of systems, the network of antennae will be able to reconstruct electronically its direction of observation or pointing, enabling it to 'see' in eight different directions each with its own independent field of view. This 'beam forming' technique means that it is no longer necessary to move the detectors or the tiles, which remain at rest: only the reconstructed beam will move. By means of long-baseline interferometry (with separations of up to 3,000 km), the instrument will achieve typical angular resolutions of 10 milliarcseconds at the 21-cm wavelength.

The international community pursuing this project numbers more than fifteen nations, among them the United States, Australia, Canada, China, India, South Africa and some European countries, including France. Beginning its operations in 2020, the instrument will have a unique impact on the observation of the earliest galaxies and stars, the determination of the nature of dark energy, the study of the re-ionization epoch, and the investigation of the end of the dark age of the universe. The SKA will, by observing large numbers of pulsars, be able to explore the strong-field gravity responsible for the merger of black holes, detecting the gravitational waves emitted. If an extraterrestrial life-form exists somewhere within our Milky Way, and is sending out radio waves, the SKA should be capable of detecting them.

Thanks to the SKA, we look forward to major scientific advances in cosmology and extragalactic astrophysics: today's less sensitive instruments can map a 'standard' spiral galaxy in the 21-cm HI line only as far out as redshift $z = 0.03$. SKA will easily reach out to $z = 2$ and we shall be able to measure the profiles of the HI lines of galaxies out to $z \sim 6$. This will give access to the distances, velocity fields and rotational curves of millions of galaxies. The much increased statistical base will allow for a greater insight into the nature of dark energy. SKA will be a world-beater, and will be able to measure profiles of the HI lines of a billion galaxies, in all parts of the sky, out to $z = 2$. This is equivalent to spectroscopic measurements of a million objects a day for three years! The construction of the SKA network should begin around the year 2012, and in 2013-14, about 10 percent of the final array could have been completed, which means that scientific programs can be under way in 2014. The whole array should be operational by 2020.

A precursor of the SKA, LOFAR (LOw Frequency ARray) is already being built in the Netherlands. This too is a new-generation radio telescope, working at low frequencies (or wavelengths greater than 1.2 m). The cost of the instrument lies mostly in its electronics rather than in the antennae, comprising a large number of fixed units (25,000 over 350 km). The signals received will be rapidly correlated and processed for observation in the desired direction. The rate of reception will be several terabits/s, and the calculating capacity tens of teraflops. The epoch of the re-ionization of the universe will be within the reach of this instrument, of a size corresponding to 10 percent of the operating surface of the SKA. Results should be coming in by 2010.

In conclusion

Astronomy has seen considerable advances in recent years. Cosmology and the formation of structures, which have long been domains of speculation, almost metaphysical at times, have today become exact sciences. We can study directly the formation and evolution of galaxies, reaching back in time with powerful telescopes.

We are beginning to comprehend the contents of our universe. However, most of it remains mysterious, and it would appear that we understand with certainty less than one percent of the total. The birth of the universe is a true laboratory of fundamental physics, in which we can experiment with particles possessing energies unlike any found in our Earthbound accelerators.

Are we at a turning point in fundamental physics, a watershed of science, which advances slowly but surely, with now and then revolutions in thought, followed by periods of the consolidation of ideas? Are we to question the law of gravity according to Newton and Einstein, in order to solve the problems of dark matter and dark energy. Do these components actually exist?

In the decades to come, the instruments deployed by astronomers and physicists will give us the answer. We are living through a formidable time!

Glossary

AGN Active galactic nuclei.

ALMA (Atacama Large Millimeter/submillimeter Array) A giant array of 12-m antennas, with baselines of several kilometers, working at submillimeter and millimeter wavelengths (from 0.3 to 3 mm), now under construction on the Altiplano de Chajnantor in the Atacama Desert of Chile (http://almaobservatory.org/, http://www.eso.org/scifacilities/alma/ and Chapter 6).

Astronomical Unit (AU) The average distance between the Earth and the Sun (149.6 million kilometers).

CDM (Cold Dark Matter) A model involving non-baryonic dark matter, made of non-relativistic particles when they decoupled from plasma at the beginning of the universe, not long after the Big Bang. It is said that some particles remained coupled with the initial 'soup,' if they could collide in reasonable numbers with the rest of the particles, and especially with their anti-particles to give photons. For this to occur, their effective cross section of collision σ had to be large enough so that the collision time ($\tau = 1/n\ \sigma v$) is less than the age of the universe. Initially, the densities (n) were so great that all the particles were in thermal equilibrium. But soon afterwards the densities decreased with expansion, and the particles decoupled. In the first seconds, the temperature of the universe was so high that particles and anti-particles of mass m were relativistic and in equilibrium with the thermal photons, whose energy hv was greater than the mass energy of the particles mc^2. The number of these particles was variable since they could be spontaneously created (at the same time as their anti-particles), from photons. When, as a result of expansion, the temperature T fell below $kT \sim mc^2$, the particles became non-relativistic and their number then decreased as $\exp(-mc^2/kT)$. It was after this epoch that the decoupling involving cold particles occurred. These did not collide with their anti-particles and their number became fixed per unit of comoving volume (except for a small fraction that disintegrated or still annihilated each other).

COBE (COsmic Background Explorer) Launched by NASA in 1989 to study the cosmic microwave background, COBE was the first orbiting telescope to discover that the fluctuations that gave rise to large structures and galaxies were only, on large scales, of the order of 10^{-5} at the time of the recombination of the universe, and were insufficient to explain the existence of today's galaxies. It

176 Mysteries of Galaxy Formation

therefore became necessary to imagine the role that non-baryonic dark matter might play (http://lambda.nasa.gov/product/cobe/).

Comoving frame of reference System within which distances between objects are measured with reference to a 'ruler' that stretches in concert with expansion. The length of this ruler is zero at the time of the Big Bang, and 1 at the present time, as the characteristic size of the (dimensionless) universe R(t).

Cosmic background radiation At a given wavelength, the totality of all photons emitted by all celestial bodies, forming a diffuse background of radiation. One aspect of this is the cosmic infrared background, which comes mainly from bright galaxies that had starburst activity in the past, and whose radiation comes to us reddened by expansion. The 'brightest' background is at microwave wavelengths, and is the radiation of the fossil black body, the vestige of the Big Bang.

Downsizing An observed effect revealing that the largest galaxies were formed very early on in the universe (at least, their stars are the oldest), and that the youngest actively star-forming galaxies today are the smallest. This phenomenon is also a feature of supermassive black holes at the centers of galaxies. The largest seem to have formed long ago, and the period when luminous quasars were numerous is past: today, only smaller galactic nuclei are active.

Eddington limit The maximum luminosity that a celestial body can have: beyond this value, radiation pressure overcomes gravity and constituents in the neighborhood of the body are ejected. Sir Arthur Eddington was an English astrophysicist who, early in the twentieth century, determined this limit for celestial bodies. This limit also applies in the case of black holes.

ELT (Extremely Large Telescopes) A new generation of ground-based telescopes, working at optical and infrared wavelengths. The European E-ELT project involves a 42-metre telescope (http://www.eso.org/ and Chapter 6), and the American counterpart is the TMT (Thirty Meter Telescope).

ERO (Extremely Red Object) Galaxies singled out by their extremely red color, in the study of very remote galaxies, or starburst activity very reddened by dust. It may well be that there are, in this class, objects with very old stellar populations (Chapter 2).

ESO (European Southern Observatory) Facility operated in the Chilean Andes by several European countries since 1962, to observe the southern sky. ESO headquarters is at Garching, near Munich in Germany (http://www.eso.org/).

GALEX (Galaxy Evolution Explorer) A satellite observing in the ultraviolet, launched by NASA in 2003, with a projected lifetime of three years (http://www.galex.caltech.edu/).

Globular clusters Very compact clusters of stars containing about 100,000 members in a spherical volume with a radius of a few parsecs. Our Galaxy has about 150 of these, and they are very old. However, they can be younger in other galaxies, especially if they were formed during galaxy mergers when starburst activity was triggered.

Gravitational lens The gravity of all massive celestial bodies causes light rays passing in their vicinity to deviate, thereby distorting the images of objects situated behind them. The focusing of the light rays causes a magnification similar to that produced by a convergent optical lens. So a distant galaxy can, if it lies in the same direction as a closer 'lens,' appear as a double or multiple image: hence the term, 'gravitational mirage.'

HDM (Hot Dark Matter) A model involving non-baryonic dark matter, made of relativistic particles when they decoupled from plasma at the beginning of the universe, not long after the Big Bang. Includes neutrinos. See **CDM**.

HST (Hubble Space Telescope) A 2.4-meter space telescope operating at optical and infrared wavelengths. Launched by NASA in April 1990, in collaboration with the European Space Agency (ESA), and most recently serviced and upgraded in May 2009 (http://hubblesite.org/).

Inflation Period of exponential expansion of the universe, just after the Big Bang, at approximately 10^{-35} seconds. The characteristic size of the universe then grew by an enormous factor, of the order of 10^{80}. Inflation was proposed by Alan Guth in 1981 in order to resolve problems of the horizon and flatness of the universe.

ISW (Integrated Sachs-Wolfe effect) A second-order effect of the cosmic microwave background, involving the blue-shifting of photons as they pass through large structures; added to first-order effects, which trace the acoustic oscillations of photons and baryons before the recombination (Chapter 5).

JWST (James Webb Space Telescope) A future space telescope, with a 6.5-meter folding segmented mirror, a collaboration between NASA and the ESA, which will replace the Hubble Space Telescope (http://www.jwst.nasa.gov/ and Chapter 6).

Kiloparsec Unit of distance equal to 1,000 parsecs.

178 Mysteries of Galaxy Formation

ΛCDM A model of the universe based on cold, non-baryonic dark matter (see **CDM**) and on a dark-energy content, of about $\Omega_\Lambda = 75\%$, making the curvature of the universe zero (i.e. a flat universe). This is now called the 'standard' model.

LBG (Lyman Break Galaxies) Distant galaxies singled out by the break in their spectra at around the continuous Lyman frequency (the threshold frequency of ionization of the hydrogen atom, 13.6 eV or a wavelength of 912 Å). The continuum from the stars is absorbed by the gas of the galaxy itself and by intergalactic gas along the line of sight. This signature enables us to study high-redshift galaxies simply by photometric means, without even obtaining a spectrum.

LOFAR (LOw Frequency ARray) Network of telescopes working at very low frequencies (wavelengths longer than 1.2 m). One of the first instruments to be used to observe the 21-cm signature of the epoch of the re-ionization of the universe (http://www.lofar.org/ and Chapter 6).

Lyman-α Spectral line of the recombination of the hydrogen atom, linking the ground state (n = 1) to the first electronic level (n = 2). The line is at 1,216 Å.

Megaparsec (Mpc) Unit of distance equal to 10^6 parsecs.

MOND (MOdified Newtonian Dynamics) A model involving modified gravity, replacing the hypothesis of non-baryonic dark matter, and attempting to explain the rotation curves of galaxies, and the formation of large-scale structures in the early universe (http://www.astro.umd.edu/~ssm/mond/ and Chapter 6).

NLS1 (Narrow-Line Seyfert 1) Seyfert galaxies are the active nuclei of galaxies, and are less energetic than quasars. There are two types: Seyfert 1 (with very wide emission lines) and Seyfert 2 (with narrower lines). In the standard unified model, Seyfert 2 galaxies would be equivalent to Seyfert 1 galaxies, but obscuration of the nucleus by dust would not allow us to observe the very wide lines close to the nucleus. NLS1s could be nuclei in the process of acquiring mass, with lines that are still narrow (Chapter 3).

PAHs (Polycyclic Aromatic Hydrocarbons) Large polycyclic aromatic molecules containing carbon and hydrogen, or small dust grains.

Parsec Unit of distance equal to 3.26 light years (or 3.08×10^{16} m). This is the distance from which the Earth-Sun distance (or Astronomical Unit) would subtend an angle of one second of arc.

QSO (Quasi-Stellar Object) or **quasar**. A term for the active nucleus of a galaxy.

Quintessence The fifth element, the existence of which is presumed in order to resolve the dark energy problem. The first four elements are baryons (neutrons and protons), leptons (especially neutrinos, an independent and massive component), photons (which are massless) and the supposed non-baryonic matter. A peculiar property of quintessence is that it exerts negative pressure, to the extent that it accelerates the expansion of the universe, instead of 'braking' it with its density.

Redshift is due to the expansion of the universe, and is often interpreted as a Doppler effect caused by galaxies receding from us with the expansion. In reality, it is the wavelength of the photons that is stretched out in the same proportions as the expansion of the characteristic scale of the universe R(t). The galaxies are in fact immobile, except that they are perturbed by local disorganized motions.

SDSS (Sloan Digital Sky Survey) Program of observations of a large part of the northern sky, designed to obtain spectra of a million galaxies. The SDSS was carried out using the 2.5-metre telescope at Apache Point, New Mexico, USA (http://www.sdss.org/).

SIDM (Self-Interacting Dark Matter) A dark-matter model in which the effective particle collision cross section is large (unlike the standard model) (Chapter 5).

SKA (Square Kilometer Array) A vast network of radio telescopes working at centimeter and meter wavelengths (from 1.2 cm to 2 m), with a total operating surface area of one million m^2 (http://www.skatelescope.org/ and Chapter 6).

SMBH (SuperMassive Black Holes) Supermassive black holes exist in practically all galactic nuclei. When they accrete mass, they become luminous as active galactic nuclei (AGN).

Spitzer The **Spitzer Space Telescope** (formerly **SIRTF**, the Space Infrared Telescope Facility) was launched by NASA in 2003. It was renamed in honor of astrophysicist Lyman Spitzer (http://www.spitzer.caltech.edu/spitzer).

Starburst (galaxies) Galaxies exhibiting bursts of star formation.

Tully-Fisher law This law relates rotation velocities to luminosity in spiral galaxies. Since the measured velocity is independent of distance, this relationship is a distance indicator for galaxies. It also serves as a test for dark matter (Chapter 5).

UCD (Ultra-Compact Dwarf) These dwarf galaxies were discovered in clusters of galaxies, and owe their particular nature to tidal interactions within the clusters (Chapter 4).

Mysteries of Galaxy Formation

WMAP (Wilkinson Microwave Anisotropy Probe) This spacecraft, launched in 2001 by NASA, investigated the cosmic microwave background and, in particular, measured peaks in acoustic oscillations. In 2003, early results from its first year of operation were already giving information about the main parameters of the universe (especially its flatness, age and baryonic content Ω_b). In 2006, the results of the first three years' observations by WMAP were published, confirming the parameters more accurately, especially those concerning the period of the re-ionization of the universe (http://map.gsfc.nasa.gov/).

Appendices

Lookback time and distance-luminosity for a given redshift z

Redshift z	Lookback time ($\times 10^9$ yr)	Distance-luminosity ($\times 10^9$ ly)
0.0	0.00	0.00
0.5	5.13	9.0
1.0	7.91	22
1.5	9.53	22
2.0	10.56	51
3.0	11.74	84
4.0	12.37	119
5.0	12.75	155
6.0	13.01	192
7.0	13.18	230
8.0	13.31	269
9.0	13.41	308
10	13.49	347
11	13.55	387
12	13.60	427
13	13.64	467
14	13.67	508
15	13.70	548
16	13.72	589
17	13.74	631
18	13.76	672
19	13.78	713
20	13.79	755
21	13.81	797
22	13.82	839
23	13.83	881
24	13.84	923
25	13.85	966
26	13.85	1008
27	13.86	1050
28	13.87	1093
29	13.87	1136
30	13.88	1178

Mysteries of Galaxy Formation

Redshift z	Lookback time (× 10⁹ yr)	Distance-luminosity (× 10⁹ ly)
32	13.89	1264
34	13.89	1350
36	13.90	1436
38	13.91	1523
40	13.91	1610
42	13.92	1696
44	13.92	1784
46	13.93	1871
48	13.93	1958
50	13.93	2046
52	13.93	2134
54	13.94	2221
56	13.94	2309
58	13.94	2398
60	13.94	2486
62	13.95	2574
64	13.95	2663
66	13.95	2751
68	13.95	2840
70	13.95	2929
72	13.95	3017
74	13.95	3106
76	13.95	3195
78	13.96	3284
80	13.96	3373
84	13.96	3552
88	13.96	3731
92	13.96	3910
96	13.96	4089

The values tabulated above are calculated for a flat universe with a Hubble constant $H_0 = 71$ km/s/Mpc, $\Omega_\Lambda = 0.72$ and $\Omega_m = 0.28$

Components of the universe

The Wilkinson Microwave Anisotropy Probe (WMAP) measured the composition of the universe. The top diagram here shows a pie chart of the relative constituents today. A similar chart (bottom) shows the composition at 380,000 years old when the cosmic background radiation which WMAP observes emanated. The composition varies as the universe expands: the dark matter and atoms become less dense as the universe expands, like an ordinary gas, but the photon and neutrino particles also lose energy as the universe expands, so their energy density decreases faster than the matter. They formed a larger fraction of the universe 13.7 billion years ago. It appears that the dark energy density does not decrease at all, so it now dominates the universe even though it was a tiny contributor 13.7 billion years ago (NASA/WMAP Science Team).

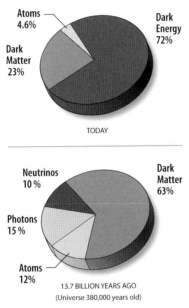

Figure A.1 WMAP data reveal that its contents include 4.6 per cent atoms, the building blocks of stars and planets. Dark matter comprises 23 per cent of the universe. This matter, different from atoms, does not emit or absorb light. It has only been detected indirectly by its gravity. The remaining 72 percent of the universe is composed of so called 'dark energy,' which acts as a sort of 'anti-gravity.' This energy, distinct from dark matter, is responsible for the present-day acceleration of the universal expansion. WMAP data is accurate to two digits, so the total of these numbers is not exactly 100 per cent. This reflects the current limits of WMAP's ability to define dark matter and dark energy. These various components of the universe are described by dimensionless quantities: their volume densities are normalized to the critical density $\rho_C = 10^{-29}$ g/cm^3. Therefore matter $\Omega_m = \rho_m/\rho_C = 0.28$ comprises ordinary matter (baryons) with $\Omega_b = \rho_m/\rho_C = 0.046$, of which only 10% is visible. The rest is non-baryonic dark matter (0.23). Finally, almost three-quarters of the universe consists of dark energy, $\Omega_\Lambda = 0.72$ or $\Lambda = 0.72$ by extension.

184 Mysteries of Galaxy Formation

The principal stages in cosmic history

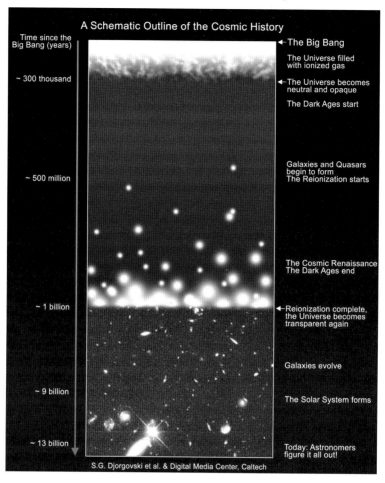

Figure A.2 The universe started with the Big Bang nearly 14 billion years ago, and from the Planck time (10^{-43} seconds) onwards its history is generally characterized by two great eras. First, there was an era during which the energy-matter content of the universe was dominated by radiation (the Radiation-Dominated Era), and second, there was the Matter-Dominated era, during which it was the material content that prevailed. Eventually, by around 380,000 years after the Big Bang, atomic nuclei and electrons had combined to make atoms of neutral gas. The glow of this 'Recombination Era' is now observed as the cosmic microwave background radiation. The universe then entered the 'Dark Ages,' which lasted for about half a billion years, until they were ended by the formation of the first galaxies and quasars. The light from these new objects turned the opaque gas filling the universe into a transparent state again, by splitting the atoms of hydrogen into free electrons and protons. This Cosmic Renaissance is also referred to by cosmologists as the 'Re-ionization Era,' and it signals the birth of the first galaxies in the early universe. (S.G. Djorgovski *et al.*, Caltech and the Caltech Digital Media Center). See also PLATE 24 in the color section.

Index

2dF survey, 95, 112
3C 75 (radio source), 83
3C 293 (radio galaxy), 154, 155
3C 273 (quasar), 73

Abell 400 (galaxy cluster), 82
Abell 1689 (galaxy cluster), 110
Abell 1835 (galaxy cluster), 64, 65
accretion disk, around black hole, 68, 72
acoustic oscillations, baryonic, 165
active galactic nuclei, 72, 73, 75, 85, 155
   energy of, 72
   mass range, 73, 75
   most luminous, 72
   very luminous, 85
adiabatic fluctuation models, 94
AGN, see active galactic nuclei
alpha elements, 32, 33
   synthesis, 32, 33, 35
alpha process, 32
Andromeda Galaxy (M31), 2, 3, 25, 27, 28, 63, 75, 116, 117, 119
   black hole at center of, 75
   distance, 3
   dwarf satellites of, 119
   mergers with, 27, 28
Anglo-Australian Telescope (AAT), 95
ångstrom, definition, 21
angular diameter distance, 8
angular momentum, in galactic disks, 139
annihilation, electron-positron, 145
   of dark matter particles, 145
astronomical unit (AU), ix, 175
Atacama Large Millimeter/submillimeter Array (ALMA), 51, 165–168, 175

Baade, Walter, 29
barred spiral galaxies, classification, 52, 54
bars, in galaxies, 106, 107
   in spiral galaxies, 52, 54, 107

baryonic matter, 86
baryonic oscillations, 124, 125
baryons, as a gas, 143
   within compact, non-radiating objects, 143
Bekenstein, Jacob, 158
bias, between visible mass and dark matter, 126
   measurement using gravitational lensing, 126
Big Bang, 3, 6, 7, 9, 11, 13, 37, 48, 51, 53, 56, 59, 61, 121, 144, 145, 184
bimodality, between red and blue galaxies, 112–113
black hole, accretion disk, 68, 72
   at center of Milky Way, 69–70
   definition, 67
   efficiency of energy conversion, 68
   mass and galactic bulge mass, relationship between, 75, 77, 91
   supermassive, nearest, 68, 71
black holes, accretion rate, 79, 89
   binary, 82, 83
      formation of, 84
      lifetimes of, 82, 84
      rotation period, 82
      search for, 84
   formation in Big Bang, 70–71
   growth of, 78–79, 89
   in early universe, 79–82
   mass of, 68, 73
   mean density, 68
   merger of, 80, 82
   mini, 70
   number in universe, 72–73, 75, 77–78
   of galactic type, 68
   of intermediate mass, 70, 80–81
   of stellar type, 68
   stellar mass, formation of, 68–69
   stellar mass, in our Galaxy, 69

supermassive, 68, 69, 71, 77–78, 155, 179
  formation of, 77–78
black-body radiation, 9, 10, 14, 48
Blanchet, Luc, 162
blue galaxies, at high redshift, 21
boson, 144
'bottom-up' model, 93, 96
brane model, of universe, 164
branes, 145
branons, 147
building blocks, of galaxies, 21–22, 37
Bullet Cluster, 110, 160, 161
Butcher-Oemler effect, 108

Canis Major dwarf galaxy, 27, 104
carbon monoxide lines, 51
CERN, 144
chain galaxy, 55, 56
Chandra X-ray Observatory, 110
chirality ('handedness'), 145
Cigar galaxy, 153
clump-clusters, 55, 56
clumpy galaxies, simulations, 57
clusters, of galaxies, 3, 4, 64, 65, 82, 96, 99,
  108–112, 116, 126, 127, 128, 137,
  160, 161
cold dark matter (CDM), 94, 96, 138, 175
  model, 122, 126, 128, 136, 137, 138, 139,
    149, 150, 151, 178
  simulation of structures, 141, 142
Coma cluster of galaxies, 96, 109
comoving distance, 8, 9
comoving frame of reference, 13, 176
composition, of universe, 182–183
concordance model, of universe, 149
core model, 137
Cosmic Background Explorer (COBE), 10,
  122, 137, 175–176
cosmic microwave background (radiation),
  5, 9, 10, 11, 12, 48, 81, 124, 125, 128,
  137, 143, 176
  acoustic oscillations in, 125
  anisotropies in, 124, 125, 128, 143
  fluctuations, 11, 12, 137
  temperature, 11, 48
    as a function of expansion, 48
Cosmic Web, 97
cosmological constant, 122
critical density, 144

cusp model, 137
Cygnus X-1, 69, 70

dark age, of the universe, 5, 6, 53, 81, 82
dark energy, 149, 163
  equation of state of, 164
dark matter, 15, 85, 86, 113, 114, 115, 121,
  122, 132–134, 138, 139, 140, 143–
  145, 147, 149, 152, 161
  around elliptical galaxies, 132–134
  baryonic, 15, 143
  cold, 94, 96, 138, 175
  concentration at centre of galaxies, 152
  distribution in dwarf galaxies, 152
  fraction in galaxies, 114
  halos, 85, 86, 113, 115
  hot, 122
  in elliptical galaxies, 134
  nature of, 143–145, 147
  non-baryonic, 15, 121, 143, 149, 161
  predicted by models, 138, 139, 140
  radial profile of, 138
decoupling, from expansion of universe,
  14, 86
deep field galaxies, detection, 59
degravitation theories, 163
density fluctuations, 10, 13, 121, 122, 123
  distribution in amplitude as function of
    scale, 122
  power spectrum as function of spatial
    frequency, 122, 123
  primordial, amplitude of, 121
detection, of remote galaxies, 47–49
diffraction limit, 49
dimensions, extra 145, 147
Doppler effect, 6, 11, 41, 71
downsizing, 85–87, 176
dust, in galaxies, 44–46
  in Milky Way, 44
  in star formation, 44
  wavelengths emitted by, 45
dwarf elliptical galaxies, see galaxies, dwarf
  elliptical
dwarf spheroidal galaxies, see galaxies,
  dwarf spheroidal
dwarfs, ultra-compact (UCDs), 116–117,
  179

earliest galaxies, search for, 37

Eddington luminosity limit, 51, 78, 79, 176
ekpyrotic model, 145
electron degeneracy pressure, 69
elliptical galaxies, 21, 33, 52, 54, 59, 60, 77,
    102, 104, 105, 108, 109, 111, 112, 113,
118, 132, 133, 136
  classification, 52, 54, 59
  fundamental plane of, 133, 136
embryonic galaxies, 15
energy distribution, in galaxy, 43–44
energy, from quasar, 69
environmental effects, on galaxy evolution,
    24, 108–109, 111–112, 113
equation of gravity, modification of, 162
escape velocity, 67
European Extremely Large Telescope (E-ELT), 169
European Southern Observatory (ESO), 176
event horizon, 67, 68
evolution, of giant galaxies, 23, 24
  of smaller galaxies, 23–24
  of stars, 68–69
expansion of universe, 6, 14
  accelerating, 7
Extremely Large Telescopes (ELTs), 176
Extremely Red Objects (EROs), 59, 176

Faber-Jackson relation, 132
Fe elements, 32–33
  formation, 32–33
feedback, negative, in massive galaxies, 116
feeding, of galactic nucleus, 49
fermion, 144
fifth force (quintessence), 163
fingers of God, 96
first structures, 9, 12
Fisher, J. Richard, 129
flattening, in dark halos, 150
Fornax cluster, 116
Freeman, Ken, 131
Freeman's law, 131
friction, dynamical, 87

GAIA astrometry satellite, 36
galactic disk sizes, 136
galaxies, accumulation of matter in, 99
  barred spiral, 52, 54
  bars in, 106–108
  beyond limiting horizon, 3
  'blue,' 112, 132, 136
    at high redshift, 21–22
  classification, 52, 54
  clumpy, 55, 56
    simulations, 57
  clusters of, 3, 4, 64, 65, 82, 96, 99,
    108–112, 116, 126, 127, 128, 137,
    160, 161
  color as function of redshift, 21
  dusty, 47
  dwarf, 116–119
    compact, 116
    elliptical, 116
    spheroidal, 116, 117, 118, 119
    tidal interactions, 117
  elliptical, 21, 33, 52, 54, 59, 60, 77, 102,
    104, 105, 108, 109, 111, 112, 113,
    118, 132, 133, 136
  fundamental plane of, 132, 133
  energy distribution, 22
  evolution, 17–21
    in clusters, 108–109, 111–112
  'evolved', 21
  faint blue, excess of, 22
  formation by monolithic collapse,
    101–102, 104, 105, 112
  formation of, 99, 101, 102, 104, 105, 106
  hierarchical formation through mergers,
    59, 85, 104, 105, 112
  high-redshift, brightness decrease, 9
    search for, 56, 58, 64
  irregular, 54, 55, 60
  lenticular, 54, 109, 136
  limiting mass, 115
  luminosity of, 112, 113
  Lyman break, 21
  massive, formation of, 87
  mass-luminosity relationship in, 131,
    132, 138
  mergers of, 22, 60, 80, 84, 85, 87, 88, 99,
    100, 101, 105, 132, 133
  orientation of, 118
  polar ring, 151
  red, 112, 132, 136
  remote, 58
  search for earliest, 37
  secular evolution of, 106–108
  Seyfert, 72, 74, 85, 88, 89

galaxies, accumulation of matter in, *cont.*
   spatial distribution of, 124
   spiral, 21, 33, 43, 46, 52, 54, 55, 60, 77,
      99, 101, 107, 108, 109, 111, 112,
      113, 130, 131, 133, 136, 156
   starburst, 39, 41–42, 43, 45, 47, 48, 49,
      152, 153, 179
   transition type, 54
   triaxial, 133
   ultra-luminous, 43, 49, 51, 58–59
   velocity dispersion of stars in, 133, 134
   with perturbed morphologies, 55
galaxy, active nucleus in, 49
   active phase of, 89
   archaeology, 29–35
   interactions, frequency in past, 55
   merger rates, evolution, 60, 62
Galaxy Evolution Explorer (GALEX), 177
General Relativity, Theory of, 8
globular star clusters, 15, 102, 103, 177
grains, in atmospheres of stars, 44
graphite, 44
gravitational lens, 177
gravitational lenses, weak, 165
gravitational lensing, 49, 50, 51, 110, 126,
     127, 136
gravitational shear, 127
gravitational telescope, 64, 65
gravitational waves, 80
gravity, modified, 156
gravothermal catastrophe, 152
'Great Attractor,' 11
great walls, of galaxies, 95, 96

halos, dark matter, 85, 86, 113, 115, 155
     simulations of, 151
   dark, of dwarf galaxies, 151–152
   massive, 115
   of galaxies, 86
   stellar, 26–27
   weak, 115
helicity, of spin, 145
Helix Nebula, 134, 135
hierarchical formation, of galaxies, 59, 85,
     104, 105
   of structures, 16–17, 24, 86, 93, 96, 119
Hills mass, 79
horizon, of universe, evolution, 6
   of visibility, personal, 6
   observable, 7
hot dark matter (HDM), 94, 122, 177
Hubble constant, 6, 8, 126
Hubble Deep Field North, 17, 19
Hubble sequence, 108, 112
Hubble Space Telescope, 17, 18, 21, 25, 38,
     39, 51, 53, 60, 73, 74, 75, 100, 101,
     103, 109, 110, 135, 155, 165, 177
Hubble time, definition, 107
Hubble tuning fork diagram, 52, 54
Hubble Ultra Deep Field, 18, 20, 51, 53, 61
Hubble, Edwin, 6, 54
Hubble's law, 6, 126
Humason, Milton, 6
hydrogen (molecular), 51
hydrogen atoms, 13

inflation theory, 122, 149, 177
Infra-Red Astronomy Satellite (IRAS), 43
Integrated Sachs-Wolfe (ISW) effect, 128,
     129, 177
interaction rate, as function of redshift, 60
inversion, of scale, 23–24
iron abundance, of star, 30
irregular galaxies, 54, 55, 60

James Webb Space Telescope (JWST), 170,
     171, 177
jets, black-hole powered, 76, 89, 90
jets, radio, 76, 82, 83, 155

Kaluza-Klein theories, 146
K-correction, negative, 47
kiloparsec, 177

Lagrange point, 10
Laplace, Pierre-Simon de
Large Hadron Collider (LHC), 144, 146, 147
large-scale structures, power spectrum of,
     149
   computer simulation of, 97
   in universe, 95
laser guide stars, 168, 169
lenticular galaxies, 54, 109, 136
Leo T dwarf spheroidal galaxy, 118
light-year, ix, 3
limiting horizon, 3, 5
Local Group, of galaxies, 25
lookback time, 8, 181–182

LOw Frequency ARray (LOFAR), 173, 178
luminosity distance, 8, 181–182
Lyman break, 21, 39, 40, 49, 64
Lyman break galaxies, 178
Lyman limit, 39
Lyman, Theodore, 38
Lyman-Alpha Blobs (LABs), 60
Lyman-α line, 37, 39, 40, 41, 42, 43, 178
   absorption by dust, 39, 41
   emission, 38, 39, 40, 42, 60, 63
   filter technique, 40
   forest, 62
   mapping projects, 40–43
M32 (galaxy), 2, 27, 29, 116
M80 (globular cluster), 103
M82 (starburst galaxy), 153
M87 (NGC 4486), 76
Madau diagram, 38, 39
Madau, Piero, 39
Magellanic Clouds, 47
mass accretion rates, 51
mass-luminosity relationship, in galaxies, 131, 132, 134
matter, baryonic, 13
   dark, 15, 85, 86, 113, 114, 115, 121, 122, 132–134, 138, 139, 140, 143–145, 147, 149, 152, 161
   non-baryonic, 15, 144
megaparsec, 178
merger rates, of galaxies, 60, 62
merger tree, 98
mergers, creating elliptical galaxies, 105, 132, 133
   involving Milky Way, 60
   of dark-matter halos, 24
   of galaxies, 22, 60, 80, 84, 85, 87, 88, 99, 100, 101, 105, 132, 133
   of small structures, 17, 24
metallicity, and age of stars, 32
   definition, 30
   of Population I stars, 30
   of Population II stars, 30
   of Population III stars, 30
   of Sun, 30
metals, definition, 30
Michell, Reverend John, 67
Milgrom, Moti, 156
Milky Way, appearance, ix
   black hole at center of, 69–70

dust in, 44
dwarf satellites of, 119
galaxy archaeology, 29–35
metallicity of stars, 32
peanut bulge, 34–35
satellite galaxies, 27
stellar halo, 27, 35–36
stellar populations, 30–31
sub-structures around, 139–142
thick disk, 34
thin disk, 33–34
tidal streams, 35–36
Millennium Simulation, 97
millimeter wave research, 49–51
millimeter waves, 47–48
missing mass, 149
modified gravity, in Solar System, 162
molecular clouds, 44
molecular lines, 51
MOND (MOdified Newtonian Dynamics) theory, 156, 157–158, 159, 161–163, 178
   and formation of galaxies, 161–162
   problem, in clusters of galaxies, 158, 161
   predictions of, 159, 161
Monoceros stream, 35
monolithic collapse, of first galaxies, 59–60

narrow-band imaging, wide-field, 40
neutralino, 144
neutrino oscillation, 144
neutrino, 'sterile,' 145, 163
neutrinos, 94, 144, 163
   flavors, 144
   'left-handed,' 163
   mass of, 144
   'right-handed,' 163
neutron capture, 33
neutron star, 69
neutrons, 13
NGC 1275 (Perseus A), 90
NGC 205, 2
non-adiabatic fluctuation models, 94
nucleosynthesis, stellar, 32–33

observable universe, 5
   edge, 3
OJ 287 (quasar), 82, 84
Olbers, Heinrich Wilhelm, 9

Olbers' paradox, 9
orbits, of stars at center of Galaxy, 70, 71
Orphan stream, 35

P Cygni profile, 42, 43
Paris Observatory, 3
parsec, ix, 178
particles, supersymmetrical, 144
Pauli pressure, 69
photons, decoupling from primitive plasma, 94
  initial domination, 15
Planck spacecraft, 137
planetary nebulae, 134
polarization, 45
polycyclic aromatic hydrocarbons (PAHs), 44–45, 46, 47, 178
Population I stars, 29, 79
Population II stars, 29, 79
Population III stars, 30, 31, 79
  collapse to black holes, 80
populations, of stars, 24
Press, William H., 17
primordial fluctuations, in universe, 11, 93, 149
protons, 13

quasar, at heart of galaxy, 69
  'cloverleaf', 51, 52
  cycle, active, 79
  energy from, 69
quasars, 49, 51, 69, 72, 73, 77–78, 81, 86, 88, 178
  first, 81
  frequency of, 73
  luminosity function of, 88
  luminosity of, 72
  number of, 77–78, 88
  redshift distribution, 86
quasi-stellar objects, see quasars
quintessence (fifth element), 179

radius, of event horizon, 68
ratio, baryons to non-baryons, 128
  dark matter to visible matter, 136
recession, of galaxies, apparent, 6
recombination era, 11, 13, 121, 126, 149, 184
redshift, 8, 21, 38, 47, 56, 179
  as Doppler effect, 6
  definition, 7
redshifts, of galaxies, 6
re-ionization era, 81, 184
re-ionization, of intergalactic medium, 40
resolution, limiting (of telescopes), 17
Rømer, Ole, 3
rotational curves, of galaxies, 156, 159
rotational velocity, of Milky Way, 156

Sagittarius A*, 69
Sagittarius dwarf galaxy, 27, 32, 35, 104
satellite halos, 139–142
scale factor r(t), 7
Schechter, Paul L., 17
Schwarzchild radius, 68
Self-Interacting Dark Matter (SIDM) model, 150, 151, 179
self-regulation, in galaxies, 89, 90, 91
Seyfert galaxies, 72, 74, 85, 88, 89, 178
shocks, heating by, 115, 150
Slipher, Vesto, 6
Sloan Digital Sky Survey (SDSS), 95, 96, 112, 123, 128, 129, 179
spectroscopy, 40, 41, 126
  long-slit, 40
speed of light, finite, 9
spin, 144
spiral galaxies, 21, 33, 43, 46, 52, 54, 55, 60, 77, 99, 101, 107, 108, 109, 111, 112, 113, 130, 131, 133, 136, 156
  classification, 52, 54
  spectra, 46
Spitzer Space Telescope, 46, 153, 179
Square Kilometer Array (SKA), 171–173, 179
stages, in cosmic history, 184
star formation, history of, in Local Group galaxies, 26
  in Milky Way, 25
  in universe, 38
  rate, evolution of, 23
  in galaxies, 59, 112, 114
starburst activity, 152, 153
  in Cigar galaxy, 153
  in galaxies, 39, 41–42, 43
starburst galaxies, 39, 41–42, 43, 45, 47, 48, 49, 152, 153, 179
  redshift in spectra, 48

stars, metal-poor, 30
  the first, 29–33
streams of stars, around Milky Way, 35, 103, 104
string theory, 163
structures, distribution in amplitude as function of scale, 122, 123
  formation of, 12–13, 14, 16, 59, 93, 96
  growth of, 13
sub-millimeter waves, 47
superclusters, of galaxies, 59
supermassive black holes (SMBH), see black holes, supermassive
SuperNova/Acceleration Probe (SNAP), 165, 166
supernovae, 30, 31, 32, 33, 44, 115, 141, 152
  Type Ia, 33, 144, 165
  Type II, 32
Supersymmetry (SUSY), 144
surface density, of galaxies, 113–114
surface, of last scattering, 5, 11, 12, 126
synchrotron radiation, 76, 82, 83

telescope, as time machine, 1
telescopes, extremely large, 168–170
The Antennae (colliding galaxies), 100
The Mice (colliding galaxies), 101
Thirty Meter Telescope (TMT), 170
'top-down' model, 93, 94
triple alpha process, 32
Tully, R. Brent, 129

Tully-Fisher relation, 129, 130, 131, 132, 136, 139, 140, 156, 157, 179
tuning fork diagram, see Hubble tuning fork diagram

ULIRGs, 58
ultra-compact dwarf (UCD) galaxies, 116–117, 179
ultra-luminous galaxies, 43, 49, 51, 58–59
universe, as gravitational lens, 8
  finite, 7
  flat, 11
  temperature decrease, 14

velocity of light, ix, 3
Very Large Telescope (VLT), 65
Virgo A, see M87 (NGC 4486)
Virgo cluster, 3, 4, 116
  distance of, 3

waves, millimeter, 9
weak interaction, 145
Weakly Interacting Massive Particles (WIMPS), 144
white dwarf, 69
Wilkinson Microwave Anisotropy Probe (WMAP), 10, 12, 14, 122, 124, 125, 128, 129, 137, 180
Williams, Robert, 19
wind, intergalactic, 83, 111

Zwicky, Fritz, 143

Printing: Mercedes-Druck, Berlin
Binding: Stein+Lehmann, Berlin